Anatomy & Physiology

an **Incredibly Easy!**® Workout

Anatomy & Physiology

an Incredibly Easy!® Workout

Wolters Kluwer | Lippincott Williams & Wilkins
Health

Philadelphia • Baltimore • New York • London
Buenos Aires • Hong Kong • Sydney • Tokyo

Staff

Executive Publisher
Judith A. Schilling McCann, RN, MSN

Editorial Director
David Moreau

Clinical Director
Joan M. Robinson, RN, MSN

Art Director
Mary Ludwicki

Senior Managing Editor
Jaime Stockslager Buss, MSPH, ELS

Clinical Project Manager
Beverly Ann Tscheschlog, RN, MS

Editor
Gale Thompson, RN, BA

Copy Editors
Kimberly Bilotta (supervisor), Dorothy P. Terry,
Pamela Wingrod

Designer
Lynn Foulk

Illustrators
Bot Roda, Judy Newhouse, Betty Winnberg

Digital Composition Services
Diane Paluba (manager), Joyce Rossi Biletz,
Donna S. Morris

Associate Manufacturing Manager
Beth J. Welsh

Editorial Assistants
Karen J. Kirk, Jeri O'Shea, Linda K. Ruhf

Workout regimen

1

The human body

The human body review

Directional terms

- *Superior* means above; *inferior* means below
- *Anterior* (or ventral) means toward the front of the body; *posterior* (or dorsal) means toward the back
- *Medial* means toward the body's midline; *lateral* means away from the body's midline
- *Proximal* means closest to the point of origin (or to the trunk); *distal* means farthest to the point of origin (or to the trunk)
- *Superficial* means toward the body surface; *deep* means farthest from the body surface

Reference planes

- Imaginary lines used to section the body and its organs; run longitudinally, horizontally, and on an angle
- Four major body reference planes: median sagittal, frontal, transverse, and oblique

Body cavities

- Dorsal—located in the posterior region of the body
- Cranial—encases the brain
- Vertebral—encloses the spinal cord
- Ventral—found in the anterior region of the trunk
- Thoracic—located superior to the abdominopelvic cavity
- Pleural—each contain a lung
- Mediastinum—contains heart, large vessels of the heart, trachea, esophagus, thymus, lymph nodes, and other blood vessels and nerves
- Abdominal—contains stomach, intestines, spleen, liver, and other organs
- Pelvic—contains bladder, some reproductive organs, and rectum
- Oral—the mouth
- Nasal—in the nose
- Orbital—house the eyes
- Middle ear—contain the small bones of the middle ear
- Synovial—enclosed within the capsules surrounding freely moveable joints

Body regions

- Umbilical
- Epigastric
- Hypogastric
- Right and left iliac (or inguinal)
- Right and left lumbar (or loin)
- Right and left hypochondriac

Cell components

- Organelles—contained in the cytoplasm and surrounded by the cell membrane
- Nucleus—responsible for cellular reproduction and division and stores deoxyribonucleic acid (DNA) (genetic material)
- Adenosine triphosphate (ATP)
- Ribosomes and endoplasmic reticulum
- Golgi apparatus
- Lysosomes

Cell reproduction

- Stage 1—mitosis (nucleus and genetic material divide)
- Stage 2—cytokineses (cytoplasm divides)

Cell division stages

- Prophase—chromosomes coil and shorten, the nuclear membrane dissolves, and chromatids connect to a centromere
- Metaphase—centromeres divide, pulling the chromosomes apart, and align in the spindle
- Anaphase—centromeres separate and pull new replicated chromosomes to the opposite sides of the cell, resulting in 46 chromosomes on each side of the cell
- Telophase—final phase; new membrane forms around 46 chromosomes through cytokinesis, producing two identical new cells

Cellular energy generation

- All cellular function depends on energy generation and transportation of substances within and among cells.
- ATP serves as the chemical fuel for cellular processes.

Movement within cells

■ Diffusion—solutes move from an area of higher concentration to one of lower concentration

■ Concentration gradient—the greater the concentration gradient the faster diffusion takes place

■ Particle size—the smaller the particles are, the faster the rate of diffusion

■ Lipid solubility—the more lipid-soluble the particles are, the more rapidly they diffuse through the lipid layers of the cell membrane

Osmosis

■ Osmosis—passive movement of fluid across a membrane from an area of lower solute concentration into an area of higher solute concentration

■ Endocytosis—an active transport method in which, instead of passing through the cell membrane, a substance is engulfed by the cell

■ Filtration—promotes transfer of fluids and dissolved materials from the blood across the capillaries into interstitial fluid

Types of tissue

■ Tissue—groups of cells that perform the same general function

■ Epithelial tissue—a continuous cellular sheet that covers the body's surface, lines body cavities, and forms certain glands

■ Loose (areolar) connective tissue—large spaces that separate the fibers and cells

■ Dense connective tissue—provides structural support and has greater fiber concentration

■ Dense regular connective tissue—consists of tightly packed fibers arranged in a consistent pattern

■ Dense irregular connective tissue—tightly packed fibers arranged in an inconsistent pattern

■ Adipose tissue —acts as insulation to conserve body heat, as a cushion for internal organs, and as a storage depot for excess food and reserve supplies of energy

■ Striated muscle tissue—striped, or striated, appearance; contracts voluntarily

■ Cardiac muscle tissue—sometimes classified as striated because it's composed of striated tissue

■ Smooth-muscle tissue—long, spindle-shaped cells; lacks the striped pattern of striated tissue

■ Neurons—highly specialized cells that generate and conduct nerve impulses

The human body can be a real puzzle, but once you understand the big picture, the small pieces fit in more easily.

■ Cross-training

When navigating the body, directional terms help you determine the exact location of a structure. Complete the following crossword puzzle to test your knowledge of these terms.

Across

1. Body cavity containing the stomach, intestines, spleen, liver, and other organs
5. Major body cavity consisting of the cranial and the vertebral cavities
7. Toward or at the body surface
9. Body cavity surrounded by the ribs and chest muscles
11. Anatomic term that means "below"
14. Slanted reference plane lying between a horizontal and a vertical plane
15. Anatomic term that means "above"
17. Toward the body's midline
18. Away from the body's midline

Down

2. Houses the heart, large vessels of the heart, trachea, esophagus, thymus, and lymph nodes
3. Body cavity inferior to the abdominal cavity
4. Farthest from the point of origin
6. Reference plane that divides the body crosswise into upper and lower halves
8. Reference plane that divides the body lengthwise into right and left regions
10. Closest to the point of origin
12. Anatomic term sometimes used instead of the term anterior
13. Cavity housing the brain
16. Body cavities containing the lungs

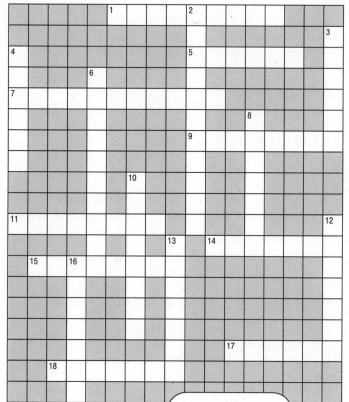

I'm more than just a bunch of viscous protoplasm, you know. My plasma membrane helps me keep my boundaries, while my nucleus controls my growth, metabolism, and reproduction.

Finish line

This cross section shows the components and structures of a cell. Label all of the components, using the clues provided.

1. Protoplasm that surrounds the nucleus _____

2. Contains digestive enzymes _____

3. Production site of adenosine triphosphate—cellular energy _____

4. Brain of the cell _____

5. Processes and packages proteins _____

6. Sites for protein synthesis _____

7. Transports protein and lipid components _____

8. Encloses the cell _____

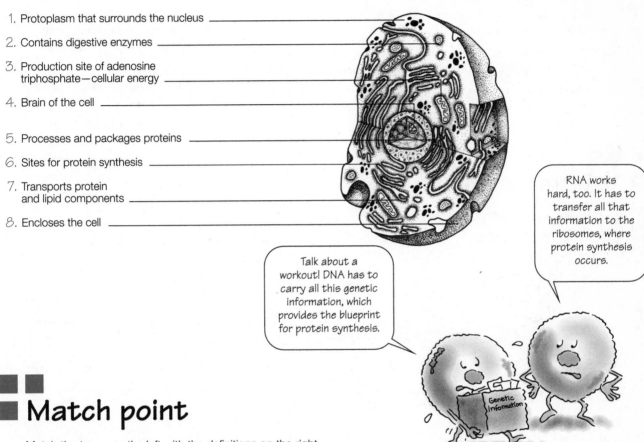

Talk about a workout! DNA has to carry all this genetic information, which provides the blueprint for protein synthesis.

RNA works hard, too. It has to transfer all that information to the ribosomes, where protein synthesis occurs.

Match point

Match the terms on the left with the definitions on the right.

1. Nucleotide _____

2. Deoxyribose _____

3. Base _____

4. Purines _____

5. Pyrimidines _____

6. Adenine _____

7. Guanine _____

8. Ribosomal RNA _____

9. Messenger RNA _____

10. Transfer RNA _____

11. Mitosis _____

12. Meiosis _____

A. Used to make ribosomes in the endoplasmic reticulum

B. Transfers the genetic code from messenger RNA for the production of a specific amino acid

C. Nitrogen-containing compound found in DNA

D. DNA base that bonds only with cytosine

E. Basic structural unit of DNA

F. Classification of double-ring compounds adenine and guanine

G. DNA base that bonds only with thymine

H. Mode of replication reserved for gametes

I. Directs the arrangement of amino acids to make proteins at the ribosomes

J. Classification of single-ring compounds thymine and cytosine

K. Five-carbon sugar found in DNA

L. Preferred mode of replication by all human cells (except gametes)

▪▪
▪ Hit or miss

Some of the following statements about cell division are true; the others are false. Mark each accordingly.

_____ 1. In mitosis, the material in the nucleus divides first, followed by division of the cell body.

_____ 2. Meiosis consists of two divisions, each resulting in two daughter cells containing a complete set of chromosomes.

_____ 3. Mitosis consists of one inactive phase and four active phases.

> **Pep talk**
>
> " The body is a marvelous machine… a chemical lab, a power-house. Every movement, voluntary or involuntary, full of secrets and marvels. "
> —Theodor Herzl

▪▪
▪ You make the call

Using the space provided, describe the processes of diffusion, osmosis, and active transport.

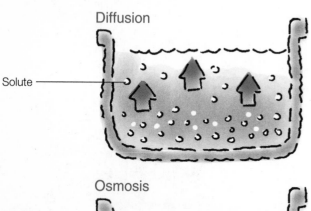

Diffusion

Solute

Osmosis

Semipermeable membrane

Solute

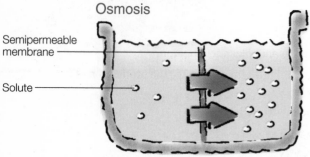

Active transport

Cell membrane

Substance

_____
_____
_____
_____
_____
_____
_____

_____
_____
_____
_____
_____
_____
_____

_____
_____
_____
_____
_____

2

Genetics and chemical organization

Genetics and chemical organization review

Genetic facts

- Genetics is the study of heredity
- Inherited traits are transmitted from parents to offspring in germ cells, or gametes
  - Human gametes are eggs (or ova) and sperm
- Inheritance is determined at fertilization

Chromosomes

- Each contains a strand of genetic material called *deoxyribonucleic acid* (DNA)
- Each trait a person inherits is carried in the genes
- Chromosomes—composed of DNA and protein
- Human ovum and sperm each contain 23 chromosomes
- When ovum and sperm unite, the result is a fertilized cell with 46 chromosomes (23 pairs) in the nucleus
- Females—have two X chromosomes
- Males—have an X and a Y chromosome
- Meiosis—cell-division process that forms ova and sperm

Genes

- Genotype—genetic information of a person
- Phenotype—recognizable traits associated with a specific genotype
- Alleles—variations of a gene (such as brown, green, or blue eye color) that can occupy a particular gene locus

Trait predominance

- Each parent contributes one set of chromosomes
- Homozygous—identical alleles on each chromosome
- Heterozygous—different alleles on each chromosome

Autosomal inheritance

- Autosomal dominant—more powerful gene and more easily expressed
- Autosomal recessive—less powerful gene and less easily expressed

Sex-linked inheritance

- The male has only one copy of most genes on the X chromosome; inheritance of those genes is called *X-linked*
- A man transmits one copy of each X-linked gene to his daughters and none to his sons

- A woman transmits one copy of each X-linked gene to each child, male or female
- Some recessive genes on X chromosomes act dominant in females

Genetic defects

- Teratogens—environmental agents that can cause congenital, structural, or functional defects

Mutations

- Permanent change in genetic material
- May cause serious defects, such as congenital anomalies

Autosomal disorders

- Single-gene disorders—error at a single gene site on DNA
- Most hereditary disorders are caused by autosomal defects
- Autosomal dominant transmission and autosomal recessive inheritance—usually affects male and female offspring equally

Sex-linked disorders

- Usually recessive traits caused by genes located on the sex chromosomes (most on X chromosome)
- Single X-linked recessive gene can cause disorder to be exhibited in a male
- All daughters of an affected male are carriers

Multifactorial disorders

- Result from a less-than-optimum expression of many different genes, not from a specific error
- Examples apparent at birth (cleft lip, cleft palate) or later (type II diabetes mellitus, hypertension)

Chromosome defects

- Congenital anomalies, or birth defects—aberrations in chromosome structure or number

Types

- Translocation—relocation of a segment of a chromosome to a nonhomologous chromosome
- Disjunction—normal separation of chromosomes during both meiosis and mitosis
- Nondisjunction—failure to separate that causes unequal distribution of chromosomes between the two resulting cells

Principles of chemistry

- Matter—anything that has mass and occupies space

- Energy—the capacity to do work (put mass into motion)
  - Potential—stored energy; kinetic—energy of motion
  - Types—chemical, electrical, and radiant

Chemical composition

- Element—matter that can't be broken down into simpler substances by normal chemical reactions
- Carbon, hydrogen, nitrogen, and oxygen account for 96% of the body's total weight

Atomic structure

- Atom—smallest unit of matter involved in chemical reactions
- Atoms of a single type constitute an element

Subatomic particles

- Protons (p+)
  - Positively charged particles
  - Each element has a distinct number of protons
- Neutrons (n)
  - Uncharged, or neutral, particles
- Electrons (e–)
  - Negatively charged particles
  - Play a key role in chemical bonds and reactions

Chemical bonds

- Force of attraction that binds a molecule's atoms together
- Formation—requires energy; breakup—releases energy

Types

- Hydrogen—two atoms associate with a hydrogen atom
- Ionic—valence electrons transfer from one atom to another
- Covalent—atoms share pairs of valence electrons

Chemical reactions

- One of two events occurs:
  1. Unpaired electrons from the outer shell of one atom transfer to the outer shell of another atom
  2. One atom shares its unpaired electrons with another atom
- Types: Synthesis, decomposition, exchange, reversible

Inorganic compounds

Water

- Body's most abundant substance
- Permits the transport of solvents
- Acts as lubricant in mucus and other bodily fluids
- Helps regulate body temperature

Electrolytes

- Compounds whose molecules consist of positively charged ions (cations) and negatively charged ions (anions)

Acids, bases, and salts

- Acids—separate into a positively charged hydrogen ion and a negatively charged anion

- Bases—separate into negatively charged hydroxide ions and positively charged cations
- Salts—form when acids react with bases
- Body fluids must attain acid-base balance for homeostasis
- More hydrogen ions = more acidic solution
- More hydroxide ions = more basic (alkaline) solution

Organic compounds

- Compounds that contain carbon or carbon-hydrogen bonds

Carbohydrates

- Sugars, starches, glycogen, and cellulose
- Release and store energy
- Three types:
  - Monosaccharides (ribose and deoxyribose)
  - Disaccharides (lactose and maltose)
  - Polysaccharides (glycogen)

Lipids

- Water-insoluble biomolecules
- Triglyceride
  - Most abundant lipid in food and the body
  - Insulates and protects
  - Body's most concentrated energy source
- Phospholipids
  - Major structural components of cell membranes
- Sterols—simple lipids with no fatty acids in their molecules
  - Bile salts—emulsify fats during digestion and aid absorption of the fat-soluble vitamins (A, D, E, K)
  - Male and female sex hormones—responsible for sexual characteristics and reproduction
  - Cholesterol—a part of animal cell membranes—needed to form all other sterols
  - Vitamin D—helps regulate calcium concentration
- Lipoproteins—help transport lipids throughout the body
- Eicosanoids
  - Prostaglandins (modify hormone responses, promote inflammatory response, open airways)
  - Leukotrienes (play role in allergic and inflammatory responses)

Proteins

- Most abundant organic compound in the body
- Composed of amino acids
- Polypeptide—many amino acids linked together
- One or more polypeptides form a protein

Nucleic acids

- Composed of nitrogenous bases, sugars, and phosphate
- DNA
  - Primary hereditary molecule
- RNA
  - Transmits genetic information from the cell nucleus to the cytoplasm
  - Guides protein synthesis from amino acids

■■ ■ Boxing match

Fill in the answers to the clues below by using all of the syllables in the box. The number of syllables for each answer is shown in parentheses. Use each syllable only once. The first answer has been provided for you as an example.

AL	CHRO	CUS	ETE	GAM	~~GE~~	GE	GE	ICS
LELES	LO	MEI	MO	MU	NET	~~NO~~	NO	NOME
O	PHE	SIS	SOME	TA	TION	~~TYPE~~	TYPE	

1. Genetic constitution of individual   (3) <u>G E N O T Y P E</u>

2. Study of heredity   (3) _ _ _ _ _ _ _ _

3. Outward manifestation of genotype   (3) _ _ _ _ _ _ _ _ _

4. Germ cell   (2) _ _ _ _ _

5. Structures of nucleus in each germ cell   (3) _ _ _ _ _ _ _ _ _ _

6. Complete set of chromosomes   (2) _ _ _ _ _ _

7. Location of gene on a chromosome   (2) _ _ _ _ _

8. Permanent change in genetic information   (3) _ _ _ _ _ _ _ _

9. Cell division process by which sperm and ova form   (3) _ _ _ _ _ _ _

10. Variations of the same gene   (2) _ _ _ _ _ _

■■ ■ Hit or miss

Some of the following statements about chromosomes are true; the others are false. Mark each accordingly.

_____ 1. A fertilized human ovum contains 23 chromosomes.

_____ 2. When an ovum and a sperm unite, corresponding chromosomes pair up.

_____ 3. Each pair of chromosomes contains information that determines a person's gender.

_____ 4. In a female, both sex chromosomes are relatively large, and each is designated by the letter X.

_____ 5. In a male, one sex chromosome is an X chromosome and one is a smaller chromosome, designated by the letter Y.

_____ 6. Each gamete produced by a male contains both an X and a Y chromosome.

_____ 7. When a sperm with an X chromosome fertilizes an ovum, the offspring may be male or female.

_____ 8. Genes are segments of a DNA chain, arranged in sequence on a chromosome.

_____ 9. The location of a specific gene on a chromosome varies from person to person.

> Some characteristics, or traits, are determined by one gene that may have many variants. Variations of the same gene are called alleles.

Coaching session
Gene expression

Gene expression refers to a gene's effect on cell structure or function. Types include:
- **Dominant:** Expressed even if only one parent passes the gene
- **Recessive:** Expressed only if both parents pass the gene
- **Codominant:** Allow equal expression of two alleles
- **Sex-linked:** Carried on sex chromosomes (most on X and are recessive; those on Y are dominant)

Strike out

The following statements pertain to genes and gene expression. Cross out all of the untrue statements.

1. Dominant genes can be expressed and transmitted to the offspring even if only one parent possesses the gene.

2. Recessive genes are expressed only under the influence of environmental factors.

3. Males have more genetic material than females.

4. Sex-linked genes are carried on sex chromosomes.

5. Almost all sex-linked genes appear on the X chromosome and are recessive.

6. Because sex-linked genes are recessive, they are rarely expressed.

7. Some genes are only expressed under certain environmental conditions.

I may look delicate, but I have a strong genotype.

Parallel bars

Complete each of the analogies below by determining the relationship between the given word pair and then determining the missing term in the second word pair.

1. Egg : _____ :: Sperm : Spermatozoa

2. Female : XX :: Male : _____

3. Inside : Genotype :: Outside : _____

4. Homozygous : Identical :: _____ : Different

5. More : _____ :: Less : Recessive

6. One too many : Trisomy :: One too few : _____

■ Jumble gym

Use the clues to help you unscramble words related to genetic defects. Then use the circled letters to answer the question posed.

Question: Which term is used to describe a disorder that results from an error at a single gene site?

1. Permanent change in genetic material

ANTIMUTO

_ ◯ _ _ _ _ _ _

2. Type of disorder resulting from a number of genes and environmental influences acting together

CAMLIATFLUTIRO

— — — — — — — — ◯ — — — —

3. Class of disorders resulting from aberrations in chromosome structure or number

ACONTINGLE SEALAMION

— — — — — — — — ◯ — — — — ◯ — — — —

4. The relocation of a segment of a chromosome to a nonhomologous chromosome

CANTORLASTION

◯ — — — — — — ◯ — — — —

5. Condition in which the number of chromosomes present is one less than normal

OMMYSOON

— — — ◯ — — —

6. The presence of an extra chromosome

MISTYOR

— — — — ◯ —

7. Occurs when chromosomes fail to separate properly during meiosis or mitosis, resulting in an unequal distribution of chromosomes between the two resulting cells

SICUNDOJINNTON

— ◯ — — — — — — — — — — —

Answer: __ __ __ ,__ __ __ __ __ __

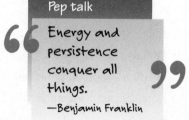

Cross-training

Complete the following crossword puzzle to test your knowledge of body chemistry.

Across

1. Matter that can't be broken down into simpler substances by normal chemical reactions

5. Dynamic equilibrium of the body

7. A form of an atom that has a different number of neutrons and, therefore, a different atomic weight

9. Compounds containing carbon

10. Formed by the combination of the atoms of two different elements

12. Most abundant organic compound in the body

13. Dense central core of an atom

14. Smallest unit of matter that can take part in a chemical reaction

15. The capacity to do work

Down

1. Negatively charged particles that orbit the nucleus in shells

2. Anything that has mass and occupies space

3. An atom's ability to combine with other atoms

4. Uncharged particles in the atom's nucleus

6. Compounds without carbon

8. Combination of two or more atoms

11. Positively charged particles in the atom's nucleus

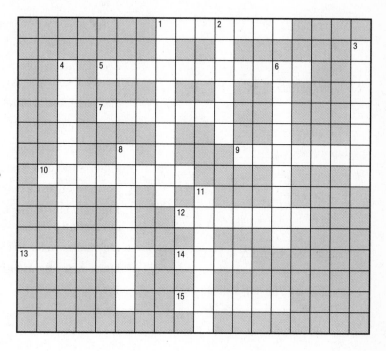

Match point

Match the terms on the left with the definitions on the right.

1. Water _____
2. Electrolytes _____
3. Salts _____
4. Carbohydrates _____
5. Triglycerides _____
6. Phospholipids _____
7. Lipoproteins _____
8. Peptide bonds _____
9. Polypeptide _____
10. Bile salts _____
11. Nucleic acid containing two long chains of deoxyribonucleotides _____
12. Nucleic acid having a single-chain structure _____
13. Chemical bond _____

A. Force of attraction that binds a molecule's atoms together
B. Help transport lipids to various parts of the body
C. Compounds whose molecules consist of positively and negatively charged ions that separate in solution
D. Main function is to release and store energy
E. Most abundant substance in the body
F. Neutral fats that insulate and protect
G. Major structural components of cell membranes
H. Form when acids react with bases
I. Emulsify fats during digestion and aid absorption of the fat-soluble vitamins
J. Link amino acids together
K. Many amino acids linked together
L. RNA
M. DNA

I think I'm having some type of reaction!

Don't worry. It's reversible. Get out of the sun and you'll feel better.

You make the call

Using the space provided, describe the four basic types of chemical reactions.

1. Synthesis reaction (anabolism)
   Hint: A + B → AB

2. Decomposition reaction (catabolism)
   Hint: AB → A + B

3. Exchange reaction
   Hint: AB + CD → A + B + C + D → AD + BC

4. Reversible reaction
   Hint: A + B ↔ AB

3

Integumentary system

Integumentary system review

Integumentary system basics

■ Includes skin and its appendages (the hair, nails, and certain glands)

Functions

Protection

■ Maintains the integrity of the body surface by migration and shedding

■ Can repair surface wounds by intensifying normal cell replacement mechanisms

■ Protects the body against noxious chemicals and invasion from bacteria and microorganisms

■ Langerhans' cells—enhance the body's immune response by helping lymphocytes to process antigens entering the skin

■ Melanocytes—protect the skin by producing the brown pigment melanin to help filter ultraviolet light

Sensory perception

■ Nerve fibers—transmit various sensations, such as temperature, touch, pressure, pain, itching—from the skin to the central nervous system

■ Autonomic nerve fibers—carry impulses to smooth muscle in the walls of the skin's blood vessels, to the muscles around the hair roots, and to the sweat glands

Body temperature regulation

■ Abundant nerves, blood vessels, and eccrine glands within the skin's deeper layer aid thermoregulation, or control of body temperature

■ When skin is exposed to cold or internal body temperature falls, blood vessels constrict, decreasing blood flow and conserving body heat

■ If skin becomes too hot or internal body temperature rises, small arteries dilate, increasing the blood flow and reducing body heat

Excretion

■ Sweat glands excrete sweat, which contains water, electrolytes, urea, and lactic acid

■ Skin
  – Eliminates body wastes through pores

  – Prevents body fluids from escaping
  – Keeps unwanted fluids in the environment from entering the body

Epidermis

■ Outermost layer of skin

■ Varies in thickness from less than 0.1 mm to more than 1 mm

■ Composed of avascular, stratified, squamous (scaly or plate-like) epithelial tissue

■ Five distinct layers (each named for its structure or function)
  – Stratum corneum—outermost layer
  – Stratum lucidum, or clear layer— blocks water penetration or loss
  – Stratum granulosum, or granular layer—responsible for keratin formation
  – Stratum spinosum, or spiny layer—helps with keratin formation and is rich in ribonucleic acid
  – Stratum basale, or the basal layer—innermost layer; produces new cells to replace those shed or worn away

Dermis

■ Also called *corium*

■ Skin's second layer

■ Contains and supports blood vessels, lymphatic vessels, nerves, and the epidermal appendages

■ Made up of extracellular material called *matrix*

■ Matrix contains:
  – collagen—a protein that gives strength to the dermis
  – elastin—makes the skin pliable
  – reticular fibers—bind the collagen and elastin fibers together

■ Two layers of dermis
  – Papillary dermis—fingerlike projections (papillae) that connect the dermis to the epidermis—contains fingerprints
  – Reticular dermis—insulates the body to conserve heat and provides energy and serves as a mechanical shock absorber

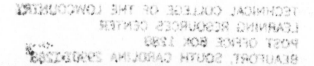

Hair

- Long, slender shafts composed of keratin
- At the expanded lower end of each hair is a bulb or root
- Root is indented by a hair papilla, a cluster of connective tissue and blood vessels
- Each hair lies within an epithelium-lined sheath called a hair follicle
  - Bundle of smooth-muscle fibers, arrector pili, extends through the dermis to attach to the base of the follicle
  - When arrector pili muscles contract, the hair stands on end
- Hair follicles also have a rich blood and nerve supply

Nails

- Situated over the distal surface of the end of each finger and toe
- Specialized types of keratin
- Nail plate—surrounded on three sides by the nail folds, or cuticles—lies on the nail bed
- Distal portion of the matrix shows through the nail as a pale crescent-moon-shaped area (lunula)
- Vascular bed imparts the characteristic pink appearance under the nails

Sebaceous glands

- Occur on all parts of the skin, except the palms and soles
- Most prominent on scalp, face, upper torso, and genitalia
- Produce sebum (a mixture of keratin, fat, and cellulose debris)
- Combined with sweat, sebum forms a moist, oily, acidic film that's mildly antibacterial and antifungal and that protects the skin surface

Sweat glands

Eccrine glands

- Widely distributed throughout the body
- Produce an odorless, watery fluid with a sodium concentration equal to that of plasma
- Those in palms and soles secrete fluid mainly in response to emotional stress
- Remaining glands respond primarily to thermal stress, effectively regulating temperature

Apocrine glands

- Located chiefly in the axillary (underarm) and anogenital (groin) areas
- Begin to function at puberty
- Body odor occurs as bacteria decompose the fluids produced by these glands

Cross-training

Test your knowledge of terms related to the integumentary system by completing the following crossword puzzle.

Across

1. Pale, crescent-moon shaped area on the nail
5. Skin's deeper layer that helps control body temperature
6. Specialized cells within the epidermis that enhance the body's immune response
11. Specific skin area supplied by sensory nerve fibers
13. Skin's top layer

Down

2. Type of sweat gland located chiefly in the axillary and anogenital areas
3. Skin cells that help protect the skin by producing brown pigment
4. Gland that produce sebum
7. Another name for subcutaneous tissue
8. Protein that gives strength to the dermis
9. Type of sweat gland producing odorless, watery fluid
10. Extracellular material that forms the dermis
12. Makes skin pliable

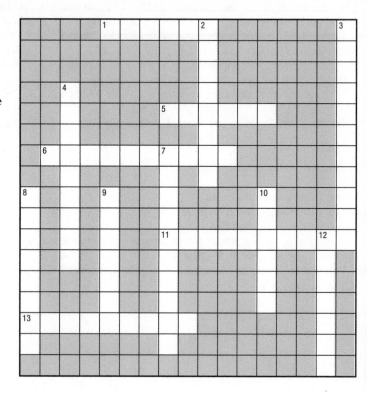

■ Power stretch

Stretch your knowledge of the integumentary system by first unscrambling the words on the left to reveal four key skin functions. Then draw lines from each box to the specific ways the skin fulfills each function.

COOTPERTIN

— — — — — — — — — —

RYESONS CREEPPOINT

— — — — — — — —
— — — — — — — — — —

BYDO PEAMUTTERER TAGLIENUOR

— — — — —
— — — — — — — — — —
— — — — — — — — —

CONREEXIT

— — — — — — — — — —

A. Maintains integrity of the body surface by migration and shedding

B. Contains nerve fibers that supply sensation to dermatomes

C. Eliminates body wastes through more than two million pores

D. Contains Langerhans' cells, which help lymphocytes process antigens entering the skin

E. Intensifies normal cell replacement mechanisms to repair surface wounds

F. Prevents the excess loss of internal body fluids by regulating the content and volume of sweat

G. Forms a layer against noxious chemicals and bacteria

H. Contains melanocytes, which produce melanin to help filter ultraviolet light

I. Has abundant nerves, blood vessels, and eccrine glands in the dermis

J. Keeps unwanted fluids in the environment from entering the body

The skin serves as the body's primary defense mechanism, protecting the body from invaders.

A-maze-ing race

Nurse Joy is leaving for her well-deserved vacation. Help her have a good time while she's gone by correctly answering questions about the skin's role in thermoregulation. Beware, though. If you answer incorrectly, Joy might get burned.

1. Joy's hike along the lava rock turned into a little more than she bargained for. Answer this question to determine where she should go next.
When internal body temperature rises, small arteries in the skin do what?
  A. Contract—Go to island 2.
  B. Dilate—Go to island 3.

2. Sorry, this exercise has gotten a little too hot for you to handle. Go back and try again.

3. Correct! As the skin becomes hot, small arteries in the second layer of skin dilate. Now answer this question to proceed.
Things are starting to heat up a bit now. Dilation of the arteries does what to blood flow?
  A. Increases it—Go to island 4.
  B. Decreases it—Go to island 5.

4. Correct! Arterial dilation increases blood flow. Go to island 6.

5. Sorry, looks like you got burned. Go back and try again.

6. Now things are *really* heating up!
If arterial dilation doesn't sufficiently reduce temperature, the eccrine glands increase production of sweat, which cools the skin via what mechanism?
  A. Transpiration—Go to island 7.
  B. Evaporation—Go to island 8.

7. Sorry, but you overheated. Go back and try again.

8. You win! You've managed to keep cool under some pretty hot circumstances.

Match point

Match each of the terms on the left with its definition on the right.

1. Subcutaneous tissue _____
2. Stratum corneum _____
3. Stratum lucidum _____
4. Stratum granulosum _____
5. Stratum spinosum _____
6. Dermis _____
7. Matrix _____
8. Collagen _____
9. Elastin _____
10. Reticular fibers _____

A. Granular epidermal layer responsible for keratin formation

B. Extracellular material that forms the dermis

C. Outermost layer of the epidermis; consists of tightly arranged layers of cellular membranes and keratin

D. Makes skin pliable

E. Elastic system that contains and supports blood vessels, nerves, and the epidermal appendages

F. Sometimes called the hypodermis

G. Clear epidermal layer that blocks water penetration or loss

H. Spiny epidermal layer that is rich in RNA and helps with keratin formation

I. Protein that gives strength to the dermis

J. Bind the collagen and elastin fibers together

It's astounding to think that something as thin as the skin can have so many layers and perform so many crucial functions.

Finish line

This cross section shows the components of the skin. Label all of the components.

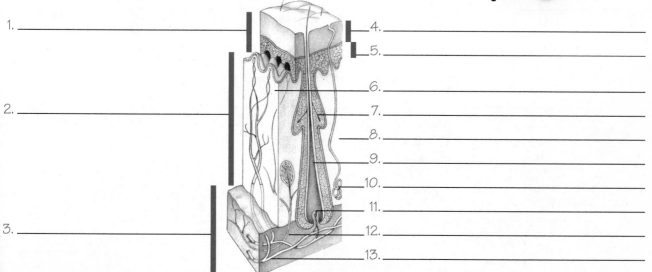

1. _____

2. _____

3. _____

4. _____

5. _____

6. _____

7. _____

8. _____

9. _____

10. _____

11. _____

12. _____

13. _____

Hit or miss

Some of the following statements about the integumentary system and its structures are true; the others are false. Mark each accordingly.

_____ 1. Hairs are composed of collagen.

_____ 2. Each hair lies within an epithelium-lined sheath called a hair follicle.

_____ 3. Nails are composed of a specialized type of bone.

_____ 4. The vascular bed imparts the characteristic pink appearance under the nails.

_____ 5. Sebaceous glands are especially plentiful on the palms and the soles.

_____ 6. Sebaceous glands produce sebum, a mixture of keratin, fat, and cellulose debris.

_____ 7. When combined with sweat, sebum has mild antibacterial and antifungal properties.

_____ 8. Hair follicles have no blood or nerve supply.

_____ 9. Eccrine glands in the palms and soles secrete fluid mainly in response to stress.

_____ 10. Apocrine glands have no known biological function.

_____ 11. Apocrine glands begin to function shortly after birth.

I think we can safely say that your apocrine glands are functioning!

Finish line

Use the illustration below to label the anatomic components of a fingernail.

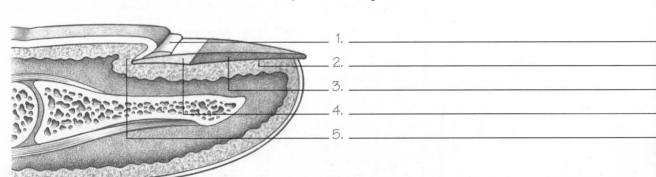

1. _____

2. _____

3. _____

4. _____

5. _____

Musculoskeletal system

Warm-up

Musculoskeletal system review

Musculoskeletal system basics

▨ Consists of muscles, tendons, ligaments, bones, cartilage, joints, and bursae
▨ Gives the human body its shape and ability to move

Body movement

▨ Musculoskeletal system works with the nervous system to produce voluntary movements
▨ Muscles contract when stimulated by impulses from the nervous system
▨ Contraction—muscle shortens—pulling on the bones to which it's attached
▨ Most movement involves groups of muscles

Muscles

▨ Three major types that are classified by the tissue they contain
  – Cardiac (heart) muscle—made up of a specialized type of striated tissue
  – Visceral (involuntary) muscle—contains smooth-muscle tissue
  – Skeletal (voluntary and reflex) muscle—consists of striated tissue; human body has about 600 skeletal muscles

Muscle functions

▨ Move body parts or the body as a whole
▨ Responsible for voluntary and reflex movements
▨ Maintain posture
▨ Generate body heat

Muscle structure

▨ Composed of large, long cell groups called muscle fibers
▨ Structures
  – Endomysium—a sheath of fibrous connective tissue that surrounds the exterior of the fiber
  – Sarcolemma—the plasma membrane of the cell that lies beneath the endomysium and just above the cells' nuclei
  – Sarcoplasm—muscle cell's cytoplasm—contained within the sarcolemma
  – Myofibrils—tiny, threadlike structures— run the fiber's length—make up the bulk of the fiber
  – Myosin (thick filaments) and actin (thin filaments)—still finer fibers within the myofibrils

▨ Perimysium—fibrous sheath of connective tissue—binds muscle fibers into a bundle, or fasciculus
▨ Epimysium—a stronger sheath—binds all of the fasciculi together to form the entire muscle

Muscle attachment

▨ Most skeletal muscles are attached to bones, either directly or indirectly
▨ Direct attachment—the epimysium of the muscle fuses to the periosteum (fibrous membrane covering the bone)
▨ Indirect attachment—the epimysium extends past the muscle as a tendon, or aponeurosis, and attaches to the bone

Moments of contraction

▨ Contraction—one bone stays relatively stationary while the other is pulled in toward the stationary one
▨ Origin—point where the muscle attaches to the stationary or less movable bone—usually lies on the proximal end of the bone
▨ Insertion—point where muscle attaches to the more movable bone—on the distal end

Muscle growth

▨ Muscle develops when existing muscle fibers hypertrophy
▨ Muscle strength and size differ among individuals because of such factors as exercise, nutrition, gender, and genetic constitution

Muscle movements

▨ Skeletal muscle permits several types of movement
▨ Muscle's functional name comes from the type of movement it permits
  – Flexor muscle permits bending (flexion)
  – Adductor muscle permits moving away from a body axis (adduction)
  – Circumductor muscle allows a circular movement (circumduction)

Muscles of the axial skeleton

▨ Essential for respiration
▨ Include:
  – muscles of the face, tongue, and neck
  – muscles of mastication
  – muscles of the vertebral column situated along the spine

Muscles of the appendicular skeleton

- Include:
  - shoulder
  - abdominopelvic cavity
  - upper and lower extremities
- Muscles of the upper extremities are classified according to the bones they move
- Muscles that move the arm are further categorized into:
  - those with an origin on the axial skeleton
  - those with an origin on the scapula

Tendons

- Bands of fibrous connective tissue
- Attach muscles to the periosteum
- Enable bones to move when skeletal muscles contract

Ligaments

- Dense, strong, flexible bands of fibrous connective tissue
- Bind bones to other bones

Bones

- Human skeleton contains 206 bones
  - 80 form the axial skeleton—called *axial* because it lies along the central line, or axis, of the body
  - 126 form the appendicular skeleton—relating to the limbs, or appendages, of the body
- Bones of the axial skeleton include:
  - facial and cranial bones
  - hyoid bone
  - vertebrae
  - ribs and sternum
- Bones of the appendicular skeleton include:
  - clavicle
  - scapula
  - humerus, radius, ulna, carpals, metacarpals, and phalanges
  - pelvic bone
  - femur, patella, fibula, tibia, tarsals, metatarsals, and phalanges

Bone classification

- Classified by shape
  - Long (such as the humerus, radius, femur, and tibia)
  - Short (such as the carpals and tarsals)
  - Flat (such as the scapula, ribs, and skull)
  - Irregular (such as the vertebrae and mandible)
  - Sesamoid (such as the patella)

Bone functions

- Protect internal tissues and organs
- Stabilize and support the body
- Provide a surface for muscle, ligament, and tendon attachment
- Move through "lever" action when contracted
- Produce red blood cells in the bone marrow (*hematopoiesis*)
- Store mineral salts

Blood supply

- Blood reaches bones through three paths:
  - Haversian canals—minute channels that lie parallel to the axis of the bone—passages for arterioles
  - Volkmann's canals—contain vessels that connect one Haversian canal to another and to the outer bone
  - Vessels in the bone ends and within the marrow

Bone formation

- At 3 months in utero, the fetal skeleton is composed of cartilage
- By about 6 months, fetal cartilage has been transformed into bony skeleton
- After birth, some bones—most notably the carpals and tarsals—ossify (harden)
- Change results from endochondral ossification—a process by which osteoblasts (bone-forming cells) produce osteoid (a collagenous material that ossifies)

Bone remodeling

- Remodeling—the continuous process whereby bone is created and destroyed
- Osteoblasts and osteoclasts—two types of osteocytes—are responsible for remodeling
- Osteoblasts—deposit new bone
- Osteoclasts— increase long-bone diameter and promote longitudinal bone growth by reabsorbing the previously deposited bone
- Epiphyseal plates—cartilage that separate the diaphysis, or shaft of a bone, from the epiphysis, or end of a bone

Cartilage

- Dense connective tissue that consists of fibers embedded in a strong, gel-like substance
- Has the flexibility of firm plastic
- Supports and shapes various structures, such as the auditory canal and the intervertebral disks
- Cushions and absorbs shock, preventing direct transmission to the bone
- Has no blood supply or innervation

- Hyaline cartilage
  - Most common type
  - Covers the articular bone surfaces—where one or more bones meet at a joint
  - Connects the ribs to the sternum
  - Appears in the trachea, bronchi, and nasal septum
- Fibrous cartilage
  - Forms the symphysis pubis and the intervertebral disks
  - Composed of small quantities of matrix and abundant fibrous elements
  - Strong and rigid
- Elastic cartilage
  - Located in the auditory canal, external ear, and epiglottis
  - Large numbers of elastic fibers give this type of cartilage elasticity and resiliency

Joints

- Also called *articulations*, points of contact between two bones that hold bones together and allow flexibility and movement
- Classified by function—extent of movement
- Classified by structure—what they're made of

Functional classification of joints

- Synarthrosis—immovable
- Amphiarthrosis—slightly movable
- Diarthrosis—freely movable

Structural classification of joints

- Fibrous joints
  - Articular surfaces of the two bones are bound closely by fibrous connective tissue
  - A little movement is possible
  - Include sutures, syndesmoses (such as the radioulnar joints), and gomphoses (such as the dental alveolar joint)
- Cartilaginous joints
  - Also called *amphiarthroses*
  - Cartilage connects one bone to another
  - Allow slight movement
  - Synchondroses—typically, temporary joints in which the intervening hyaline cartilage converts to bone by adulthood—for example, the epiphyseal plates of long bones
  - Symphyses—joints with an intervening pad of fibrocartilage—for example, the symphysis pubis
- Synovial
  - Bony surfaces in the synovial joints—separated by a viscous, lubricating fluid (the synovia) and by cartilage
  - Joined by ligaments lined with a synovia-producing membrane
  - Freely movable
  - Include most joints of the arms and legs.

- Other features of synovial joints include:
  - joint cavity—a potential space that separates the articulating surfaces of the two bones
  - articular capsule—a saclike envelope with outer layer that is lined with a vascular synovial membrane
  - reinforcing ligaments—fibrous tissue that connects bones within the joint and reinforces the joint capsule

Types of synovial joints

- Gliding
  - Flat or slightly curved articular surfaces
  - Allow gliding movements
  - May not allow movement in all directions
  - Examples: intertarsal and intercarpal joints
- Hinge
  - Convex portion of one bone fits into a concave portion of another
  - Movement resembles that of a metal hinge
  - Movement is limited to flexion and extension
  - Examples: elbow and knee
- Pivotal
  - Rounded portion of one bone fits into a groove in another bone
  - Allow only uniaxial rotation of the first bone around the second
  - Example: head of the radius—rotates within a groove of the ulna
- Condylar joints—oval surface of one bone fits into a concavity in another bone
  - Allow flexion, extension, abduction, adduction, and circumduction
  - Examples: radiocarpal and metacarpophalangeal joints of the hand
- Saddle joints
  - Resemble condylar joints but allow greater freedom of movement
  - Carpometacarpal joints of the thumb—only saddle joints in the body
- Ball-and-socket joint
  - Gets its name from the way its bones connect
  - Spherical head of one bone fits into a concave "socket" of another bone
  - Shoulder and hip joints—body's only ball-and-socket joints

Bursae

- Small synovial fluid sacs
- Located at friction points around joints between tendons, ligaments, and bones
- Act as cushions to decrease stress on adjacent structures
- Examples include:
  - subacromial bursa (located in the shoulder)
  - prepatellar bursa (located in the knee)

■ Power stretch

Stretch your knowledge of the musculoskeletal system by first unscrambling the words on the left to reveal the three major types of muscle in the human body. Then draw a line from the box to the characteristics of each muscle.

ACIDARC

— — — — — — —

OHTOMS

— — — — — —

LATKELES

— — — — — — — —

A. Responsible for voluntary and reflex movement

B. Consists of striated tissue

C. Responsible for involuntary movements

D. Found in heart tissue

E. Move body parts or the body as a whole

F. Made up of a specialized type of striated tissue

■ Hit or miss

Some of the following statements are true; the others are false. Mark each accordingly.

_____ 1. During contraction, force is applied to the tendon and the bone is moved.

_____ 2. Most movement involves the action of a single muscle.

_____ 3. The human body has about 600 muscles.

_____ 4. During muscle contraction, myosin and actin slide over each other, reducing sarcomere length.

_____ 5. When a muscle fiber is viewed microscopically, it has a smooth appearance.

_____ 6. A fibrous sheath of connective tissue, called the perimysium, binds muscle fibers into a bundle, or fasciculus.

_____ 7. The epimysium is a thin layer of connective tissue on the inside of the perimysium.

_____ 8. Extending beyond the muscle, epimysium forms the bursae.

_____ 9. Most skeletal muscles are attached to bones indirectly by way of a tendon.

_____ 10. Skeletal muscles may attach directly to bone by fusing to the periosteum.

■ Cross-training

Assess your knowledge of musculoskeletal system terms by completing the following crossword puzzle.

Across

1. Functional units of skeletal muscle
5. Movement toward a body axis
6. The muscle cell's cytoplasm, contained within the sarcolemma
8. Circular movement
10. Fibrous membrane covering a bone
12. Bone-forming cell
13. Point where muscle attaches to the stationary or less-movable bone
14. Plasma membrane of the cell (lies beneath the endomysium and just above the cell's nucleus)

Down

2. Sheath of fibrous connective tissue surrounding a muscle fiber
3. Bands of fibrous connective tissue that attach muscle to the periosteum
4. Dense, strong, flexible bands of fibrous connective tissue that bind bones to other bones
7. Point where a muscle attaches to the more movable bone
9. Tiny, threadlike structures comprising the bulk of a muscle fiber
11. Point of contact between two bones

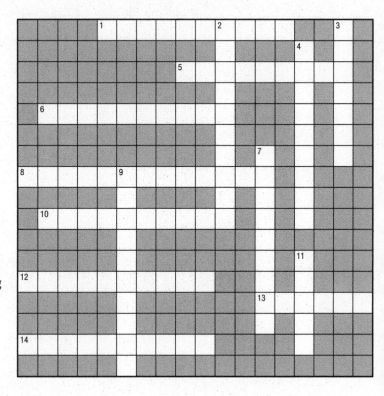

The musculoskeletal system gives me shape, allows me to move, and helps keep me warm! That's my kind of system!

■ Finish line

The following illustration shows anterior and posterior views of some of the major muscles. Flex your mental muscle, and see if you can name them all.

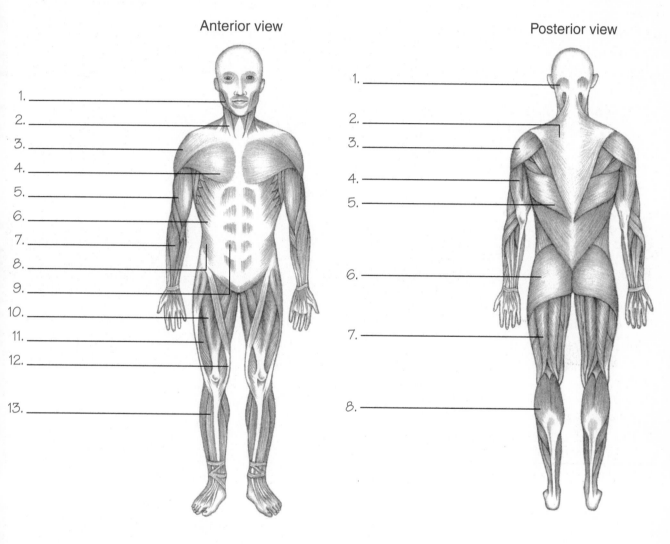

Anterior view

Posterior view

1. _____
2. _____
3. _____
4. _____
5. _____
6. _____
7. _____
8. _____
9. _____
10. _____
11. _____
12. _____
13. _____

1. _____
2. _____
3. _____
4. _____
5. _____
6. _____
7. _____
8. _____

You make the call

Muscles permit a variety of movements, as illustrated in the drawings below. Label each illustration with the type of movement being shown.

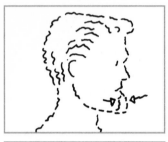

1. _____
(moving backward)

2. _____
(moving forward)

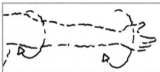

3. _____
(moving in a circular manner)

4. _____
(straightening, increasing the joint angle)

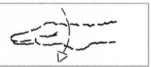

5. _____
(bending, decreasing the joint angle)

6. _____
(turning downward)

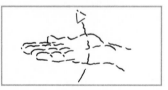

7. _____
(turning upward)

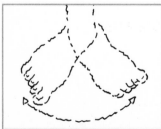

8. _____
(turning outward)

9. _____
(turning inward)

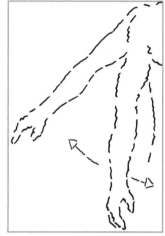

10. _____
(moving away from midline

11. _____
(moving toward the midline)

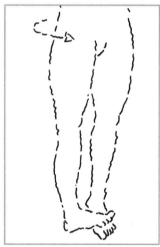

12. _____
(turning toward the midline)

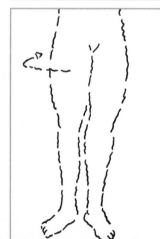

13. _____
(turning away from the midline)

■ Finish line

The illustration below shows the major bones and bone groups in the body. Test your knowledge of the skeletal system by naming them all.

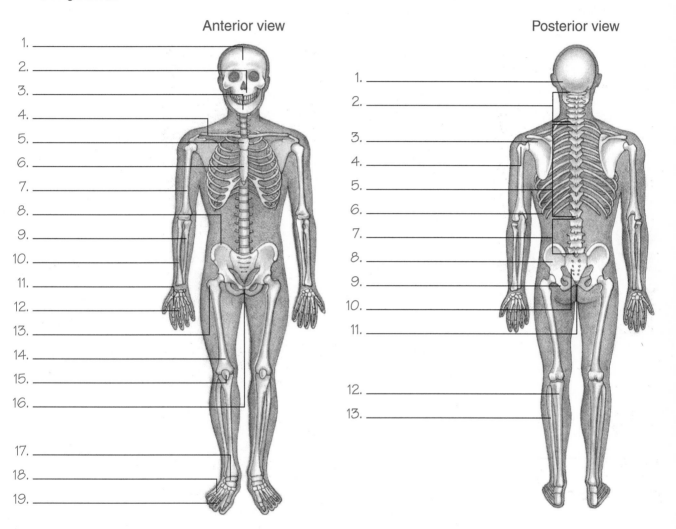

Anterior view

1. _____
2. _____
3. _____
4. _____
5. _____
6. _____
7. _____
8. _____
9. _____
10. _____
11. _____
12. _____
13. _____
14. _____
15. _____
16. _____

17. _____
18. _____
19. _____

Posterior view

1. _____
2. _____
3. _____
4. _____
5. _____
6. _____
7. _____
8. _____
9. _____
10. _____
11. _____

12. _____
13. _____

Coaching session

Types of bone tissue

Each bone consists of two layers:
• **Compact bone:** Dense, smooth outer layer with haversian canals and layers of calcified matrix
• **Cancellous bone:** Spongy inner layer made up of trabeculae and red marrow; no canals

Match point

Match the terms on the left with the definitions on the right.

1. Sarcomere _____
2. Aponeurosis _____
3. Periosteum _____
4. Origin _____
5. Insertion _____
6. Axial skeleton _____
7. Appendicular skeleton _____
8. Tendon _____
9. Ligament _____
10. Circumduction _____

A. Lies along the central line or axis of the body
B. Functional unit of skeletal muscle
C. Circular movement of a joint
D. Relates to the limbs or appendages of the body
E. Fibrous membrane covering bone
F. Binds bones to other bones
G. Location where a muscle becomes a tendon
H. Enables bones to move when skeletal muscles contract
I. Point where a muscle attaches to the more movable bone
J. Point where a muscle attaches to stationary or less movable bone

Power stretch

The human skeleton contains 206 bones, which are divided into two categories. First, unscramble the words in the boxes below to identify the names of the two skeletons. Then draw a line from each box to the pertinent bones and muscles of that skeleton.

LAXIA

— — — — —

CARPALDUEPIN

— — — — — — — — — — — —

A. Facial and cranial bones
B. Muscles of the upper and lower extremities
C. Shoulder muscles
D. Respiratory muscles
E. Hyoid bone
F. Abdominopelvic muscles
G. Vertebrae
H. Humerus, radius, and ulna
I. Pelvic bones
J. Ribs and sternum
K. Muscles of mastication
L. Clavicle
M. Muscles of the vertebral column
N. Scapula
O. Femur, patella, fibula, and tibia
P. Muscles of the face, tongue, and neck

Odd man out

The following lists of bones are grouped according to shape. Circle the odd man out in each group—in this case, the bone with a different shape. Note that one shape isn't represented here.

1. Humerus   Scapula   Tibia

2. Carpals   Radius   Tarsals

3. Rib   Scapula   Skull   Ulna

4. Femur   Mandible   Vertebra

Strike out

As an individual ages, a number of musculoskeletal changes occur. Cross out the changes below that are not commonly associated with aging.

1. Decreased height

2. Lengthening of arms

3. Decreased muscle mass

4. Accelerated collagen formation

5. Increased red blood cell production

6. Increased viscosity of synovial fluid

7. Thinning of synovial membranes

8. Shortening of the trunk

Memory jogger

To remember the functions of muscles, think "Miles Per Hour":

M_____

P _____

H_____

■ Finish line

This illustration shows the inside of a long bone.
Label the parts of the bone, as indicated.

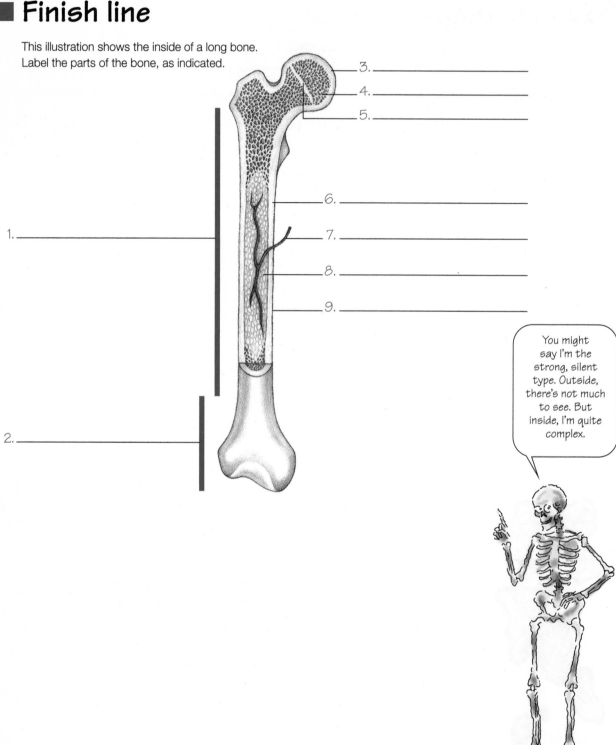

3. _____

4. _____

5. _____

6. _____

7. _____

8. _____

9. _____

1. _____

2. _____

You might say I'm the strong, silent type. Outside, there's not much to see. But inside, I'm quite complex.

Train your brain

Sound out each group of pictures and symbols to reveal information about some of the bones' functions.

Power stretch

Unscramble the words on the left to reveal two types of bone tissue. Then draw a line from the box to the characteristics of each bone type.

A. Spongy in structure

B. Consists of layers of calcified matrix containing spaces occupied by osteocytes

C. Forms inner layer of bone

D. Consists of tiny spikes, called trabeculae

E. Forms the outer layer of bone

F. Fills the central regions of the epiphyses and the inner portions of short, flat, and irregular bones

G. Found in the diaphyses of long bones and the outer layers of short, flat, and irregular bones

H. Contain haversian canals

CAMPCOT

_ _ _ _ _ _ _

CLONALCUES

_ _ _ _ _ _ _ _ _ _

■■ Cross-training

Strengthen your knowledge of terms pertaining to bones by completing the following crossword puzzle.

Across

1. Bone layers
5. The production of red blood cells
11. Point of contact between two bones
12. The ossification of cartilage into bone

Down

1. Small cavities containing osteocytes inside compact bone
2. The end of a long bone
3. Type of bone found in the diaphyses of long bones
4. Process of bone hardening
6. Bone-forming cell
7. Type of bone filling the central regions of the epiphyses
8. Dense connective tissue that consists of fibers embedded in a strong, gel-like substance
9. The shaft of a long bone
10. A collagenous material produced by osteoblasts

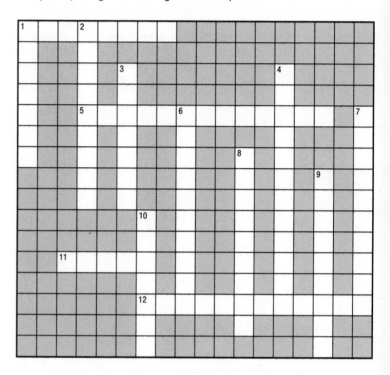

No bones about it! There's a lot to know about the musculoskeletal system.

■ Hit or miss

Some of the following statements about bone formation and remodeling are true; the others are false. Mark each accordingly.

_____ 1. At 3 months in utero, the fetal skeleton is composed of cartilage.

_____ 2. Fetal cartilage isn't transformed into bony skeleton until shortly before birth.

_____ 3. After birth, the carpals and the tarsals harden first.

_____ 4. Osteoblasts and osteoclasts are two types of osteocytes responsible for remodeling.

_____ 5. Bone growth continues until the epiphyseal plates ossify, usually by age 12.

_____ 6. Bone density begins to decrease at age 50 in both women and men.

■ Strike out

Some of the following statements about joints and cartilage are incorrect. Cross out all of the incorrect statements.

1. Cartilage is a dense connective tissue that supports and shapes various structures.

2. Cartilage has a rich blood supply.

3. Fibrous cartilage covers the articular bone surfaces and connects the ribs to the sternum.

4. Elastic cartilage is located in the auditory canal, external ear, and epiglottis.

5. Joints are classified according to the size of the bone comprising the joint.

6. Synovial joints are freely movable and include most joints of the arms and legs.

7. Soft tissue separates the contiguous bony surfaces in synovial joints.

You make the call

Following are four illustrations representing the four stages of bone growth and remodeling of the epiphyses of a long bone. Using the space provided, describe what's occurring in each stage.

Creation of an ossification center

_____

_____

_____

_____

_____

_____

_____

_____

_____

_____

_____

Epiphysis

Diaphysis

Growing hyaline cartilage

Cavities

Enlarged cartilage cells

Growing hyaline cartilage

Medullary cavity

Osteoblasts form bone

_____

_____

_____

_____

_____

_____

_____

_____

_____

_____

_____

Growing cartilage

Osteoblasts creating bone

Trabeculae

Growing cartilage

Epiphyseal plate

Medullary cavity

Bone length grows

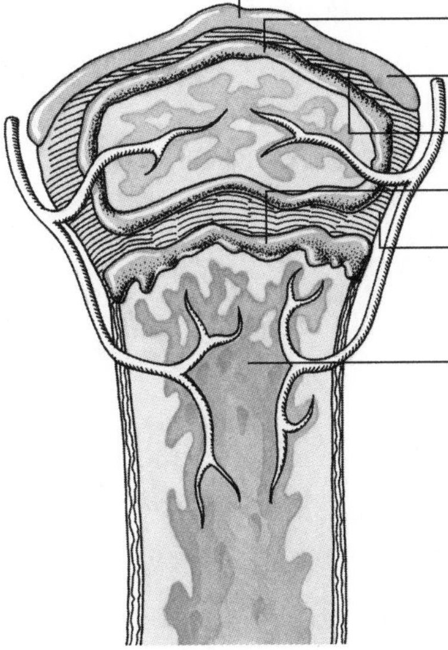

_____

_____

_____

_____

_____

_____

_____

_____

_____

_____

_____

Growing cartilage

Epiphyseal plate

Articular cartilage

Ossifying cartilage

New trabeculae

Compact bone replacing cartilage

Medullary cavity

Remodeling

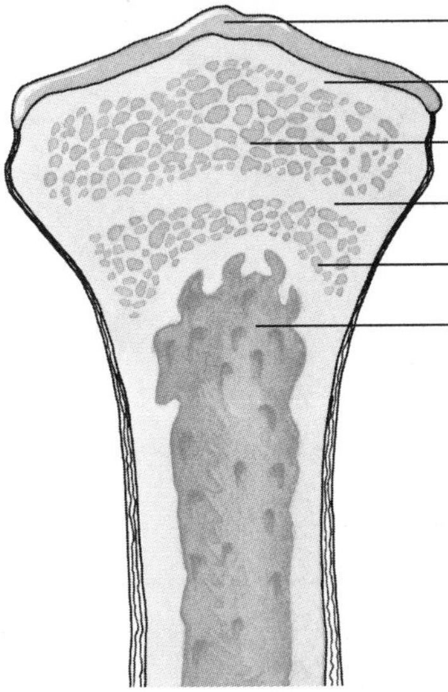

_____

_____

_____

_____

_____

_____

_____

_____

_____

_____

_____

_____

Articular cartilage

Compact bone

Cancellous bone

Epiphyseal line

Cancellous bone

Medullary cavity

Train your brain

Sound out each group of pictures and symbols to reveal information about bone development.

1.

2.

Match point

Match these musculoskeletal structures with their definitions.

1. Hyaline cartilage _____
2. Fibrous cartilage _____
3. Elastic cartilage _____
4. Joints _____
5. Fibrous joints _____
6. Cartilaginous joints _____
7. Symphysis _____
8. Synchondroses _____
9. Synovial joints _____
10. Gliding joints _____
11. Hinge joints _____
12. Pivot joint _____
13. Condylar joint _____
14. Saddle joint _____
15. Ball-and-socket joint _____
16. Bursae _____

A. Joint with an intervening pad of fibrocartilage

B. Most common type of cartilage

C. Joints consisting of two bones joined together by cartilage

D. Joint in which the rounded portion of one bone fits into the groove of another bone

E. Joint in which the spherical head of one bone fits into a concave "socket" of another bone

F. Points of contact between two bones that hold the bones together

G. Joint that's similar to but allows greater movement than a condylar joint

H. Joints joined by ligaments lined with a synovia-producing membrane

I. Joints in which a convex portion of one bone fits into a concave portion of another

J. Temporary joints in which hyaline cartilage converts to bone by adulthood

K. Joints with flat or slightly curved articular surfaces that allow gliding movements

L. Cartilage containing large numbers of elastic fibers, making it elastic and resilient

M. Joint in which an oval surface of one bone fits into a concavity in another bone

N. Small synovial fluid sacs located at friction points around joints between tendons, ligaments, and bones

O. Joints consisting of two bones bound closely together at articular surfaces by fibrous connective tissue, making little movement possible

P. Cartilage composed of small quantities of matrix and abundant fibrous elements, making it strong and rigid

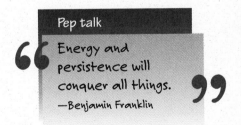

Pep talk

Energy and persistence will conquer all things.
—Benjamin Franklin

■ Match point

Match the joint on the left with its location, listed on the right.

1. Gliding joint _____
2. Hinge joints _____
3. Pivot joint _____
4. Condylar joints _____
5. Ball-and-socket joints _____
6. Saddle joints _____

A. Intertarsal and intercarpal joints of the hands and feet
B. Carpometacarpal joints of the thumb
C. Radius
D. Elbow and knee
E. Shoulder and hip joints
F. Radiocarpal and metacarpophalangeal joints of the hand

■ Jumble gym

Unscramble the first three words below to discover the terms used to classify a joint by function; then unscramble the next three words to discover the terms used to classify a joint by structure. Finally, use the circled letters to answer the question posed.

Question: **Which term is another word that can be used to refer to the joints?**

1.
A S H S I R N O R S T Y _ _ _ _ _ ◯ _ _ _ _ ◯ _

2.
H A R P S A M H I T S I O R _ _ _ _ _ ◯ _ _ _ _ _ ◯ _ _

3.
R A D I O S H I R T S _ _ _ _ ◯ _ ◯ _ _ _ _

4.
B I F O U R S _ ◯ _ _ _ ◯ _

5.
A I L S T A I N C U R G O ◯ _ _ _ _ _ ◯ _ _ ◯ _ _ _

6.
N A V Y S O I L _ _ _ ◯ _ _ _ ◯

Answer: _ _ _ _ _ _ _ _ _ _ _ _ _ _ _

Neurosensory system

Neurosensory system review

Nervous system basics

- The nervous system coordinates all body functions
- Has two main types of cells:
  - Neurons—conducting cells
  - Neuroglia—supportive cells

Neurons

- Basic unit of the nervous system
- One axon and many dendrites extend from the central cell body and transmit signals
- Responsible for neurotransmission—conduction of electro-chemical impulses throughout the nervous system
- Neuron activity may be provoked by mechanical stimuli, thermal stimuli, or chemical stimuli

Axons

- Conducts nerve impulses away from the cell body
- Wrapped in a white, fatty covering called a myelin sheath
- Myelin sheath is produced by Schwann cells—phagocytic cells separated by gaps called nodes of Ranvier

Dendrites

- Short, thick, diffusely branched extensions of the cell body
- Receive impulses from other cells
- Conduct impulses toward the cell body

Neuroglia

- Also called *glial cells*
- Supportive cells of the nervous system
- Form roughly 40% of the brain's bulk
- Four types
  - Astroglia, or astrocytes—supply nutrients to neurons and help them maintain their electrical potential as well as form part of the blood-brain barrier
  - Ependymal cells—line the brain ventricles and choroid plexuses; help produce cerebrospinal fluid (CSF)
  - Microglia—phagocytic cells that ingest and digest microorganisms and waste products from injured neurons
  - Oligodendroglia—support and electrically insulate central nervous system axons by forming protective myelin sheaths

Central nervous system

- Includes the brain and the spinal cord
- Encased by the bones of the skull and vertebral column
- Protected by CSF and the meninges

Brain

- Consists of cerebrum, cerebellum, brain stem, diencephalon, limbic system, and reticular activating system

CEREBRUM

- Largest part of brain
- Houses the nerve center that controls sensory and motor activities and intelligence
- Outer cortex of gray matter (consisting of unmyelinated nerve fibers); inner layer of nerve fibers
- White matter (myelinated nerve fibers within the cerebrum)
- Basal ganglia—control motor coordination and steadiness
- Has right and left hemispheres, each divided into four lobes—frontal, temporal, parietal, and occipital
- Corpus callosum bridges the hemispheres and allows communication between corresponding centers in each hemisphere
- Rolling surface of the cerebrum is made up of gyri (convolutions) and sulci (creases or fissures)

CEREBELLUM

- Brain's second largest region
- Lies behind and below the cerebrum
- Has two hemispheres
- Outer cortex of gray matter; inner core of white matter
- Functions to maintain muscle tone, coordinate muscle movement, and control balance

BRAIN STEM

- Lies below the cerebrum, in front of the cerebellum
- Consists of the midbrain, pons, and medulla oblongata
- Relays messages between the parts of the nervous system
- Three main functions
  1. Produces autonomic behaviors necessary for survival
  2. Provides pathways for nerve fibers between higher and lower neural centers
  3. Serves as the origin for 10 of the 12 pairs of cranial nerves

MIDBRAIN

- Reflex center for cranial nerves III and IV
- Mediates pupillary reflexes and eye movements

PONS

- Helps regulate respirations
- Connects the cerebellum with the cerebrum
- Links the midbrain to the medulla oblongata
- Reflex center for cranial nerves V through VIII
- Mediates chewing, taste, hearing, and equilibrium

MEDULLA OBLONGATA

- Joins the spinal cord at the level of the foramen magnum
- Influences cardiac, respiratory, and vasomotor functions
- Center for the vomiting, coughing, and hiccuping reflexes

DIENCEPHALON

- Located between the cerebrum and the midbrain
- Consists of the thalamus and hypothalamus

THALAMUS

- Relays sensory stimuli (except olfactory) to cerebral cortex
- Functions include primitive awareness of pain, screening of incoming stimuli, and focusing of attention

HYPOTHALAMUS

- Controls or affects body temperature, appetite, water balance, pituitary secretions, emotions, and autonomic functions, including sleeping and waking cycles

LIMBIC SYSTEM

- Primitive brain area deep within the temporal lobe
- Initiates basic drives
- Screens all sensory messages traveling to the cerebral cortex

RETICULAR ACTIVATING SYSTEM

- Diffuse network of hyperexcitable neurons
- Fans out from the brain stem through the cerebral cortex
- Functions as the alert system for the cerebral cortex

Oxygenating the brain

- Four major arteries supply the brain with oxygenated blood
- Basilar artery supplies oxygen to the posterior brain
- Common carotids branch into the two internal carotids, which divide further to supply oxygen to the anterior brain and the middle brain
- Arteries interconnect through the circle of Willis
- Circle of Willis ensures that oxygen is continually circulated to the brain

Spinal cord

- Cylindrical structure in the vertebral canal
- Extends from the foramen magnum at the base of the skull to the upper lumbar region of the vertebral column
- Give rise to spinal nerves
- At inferior end, nerve roots cluster in the cauda equina
- H-shaped mass of gray matter is divided into horns
- Cell bodies in the two dorsal horns primarily relay sensations
- Cell bodies in the two ventral horns play a part in voluntary and reflex motor activity
- White matter surrounds the horns

SENSORY PATHWAYS

- Sensory impulses travel via the afferent (sensory, or ascending) neural pathways to the sensory cortex
- Impulses use two major pathways—dorsal horn and ganglia

DORSAL HORN

- Pain and temperature sensations enter the spinal cord through the dorsal horn and then travel to the thalamus

GANGLIA

- Touch, pressure, and vibration sensations enter the cord via relay stations called ganglia
- Impulses travel up the cord and enter the thalamus
- Thalamus relays all incoming sensory impulses (except olfactory impulses) to the sensory cortex for interpretation

MOTOR PATHWAYS

- Impulses travel from the brain to the muscles via the efferent (motor, or descending) neural pathways
- Upper motor neurons originate in the brain and form two major systems:
  - Pyramidal system—responsible for fine, skilled movements of skeletal muscle
  - Extrapyramidal system—controls gross motor movements

REFLEX RESPONSES

- Occur automatically to protect the body
- Spinal nerves—mediate deep tendon reflexes, superficial reflexes and, in infants, primitive reflexes

DEEP TENDON REFLEXES

- Biceps reflex—forces flexion of the forearm
- Triceps reflex—forces extension of the forearm
- Brachioradialis reflex—causes supination of the hand and flexion of the forearm at the elbow
- Patellar reflex—forces contraction of the quadriceps muscle in the thigh with extension of the leg
- Achilles reflex—forces plantar flexion of the foot at the ankle

SUPERFICIAL REFLEXES

- Reflexes of skin and mucous membranes
- Examples: plantar flexion of toes, Babinski's response, cremasteric reflex, abdominal reflex

PRIMITIVE REFLEXES

- Abnormal in adults but normal in infants
- Disappear as the neurologic system matures
- Examples: grasping reflex, sucking reflex, glabella reflex

Protective structures

- Brain and spinal cord are protected from shock and infection by the bony skull and vertebrae, CSF, and three membranes: the dura mater, arachnoid membrane, and pia mater

DURA MATER

- Tough, fibrous, leather-like tissue composed of two layers
- Endosteal dura—forms the periosteum of the skull
- Meningeal dura—provides support and protection

ARACHNOID MEMBRANE

▪ Thin, fibrous membrane that hugs the brain and spinal cord

PIA MATER

▪ Continuous, delicate layer of connective tissue that covers and contours the spinal tissue and brain
▪ Subdural space—between dura mater and arachnoid membrane
▪ Subarachnoid space—between pia mater and arachnoid membrane
▪ CSF—protects brain and spinal tissue from jolts and blows

Peripheral nervous system

▪ Consists of the cranial nerves, spinal nerves, and autonomic nervous system (ANS)

Cranial nerves

▪ 12 pairs of cranial nerves
▪ Transmit motor or sensory messages (or both) primarily between the brain or brain stem and the head and neck
▪ All except the olfactory and optic nerves exit from the midbrain, pons, or medulla oblongata

Spinal nerves

▪ Each of the 31 pairs is named for the vertebra immediately below the nerve's exit point from the spinal cord
▪ Designated as C1 through S5 and the coccygeal nerve
▪ Each consists of afferent (sensory) and efferent (motor) neurons
▪ Each carry messages to and from particular body regions, called dermatomes

Autonomic nervous system

▪ Innervates (supplies nerves to) all internal organs
▪ Nerves of the ANS carry messages to the viscera from the brain stem and neuroendocrine regulatory centers
▪ Dual innervation allows two divisions to counterbalance each other

SYMPATHETIC NERVOUS SYSTEM

▪ Produces widespread, generalized physiologic responses
 – Vasoconstriction
 – Elevated blood pressure
 – Enhanced blood flow to skeletal muscles
 – Increased heart rate and contractility
 – Increased respiratory rate
 – Smooth-muscle relaxation of the bronchioles, GI tract, and urinary tract
 – Sphincter contraction
 – Pupillary dilation and ciliary muscle relaxation
 – Increased sweat gland secretion
 – Reduced pancreatic secretion

PARASYMPATHETIC NERVOUS SYSTEM

▪ Creates specific responses involving only one organ or gland such as:

 – reductions in heart rate, contractility, and conduction velocity
 – bronchial smooth-muscle constriction
 – increased GI tract tone and peristalsis, with sphincter relaxation
 – increased bladder tone and urinary system sphincter relaxation
 – vasodilation of external genitalia, causing erection
 – pupil constriction
 – increased pancreatic, salivary, and lacrimal secretions

Special sense organs

Eye

▪ Organ of vision
▪ Contains about 70% of the body's sensory receptors

EXTRAOCULAR EYE STRUCTURES

▪ Extraocular muscles hold the eyes in place and control their movement
▪ As one muscle contracts, its opposing muscle relaxes
▪ Includes eyelids, conjunctivae, and lacrimal apparatus
▪ Support and protect the eyeball

EYELIDS (PALPEBRAE)

▪ Loose folds of skin that cover the anterior portion of the eye
▪ Lid margins contain eyelashes and sebaceous glands
▪ Contain three types of glands:
 – meibomian glands—secrete sebum
 – glands of Zeis—connected to the follicles of the eyelashes
 – Moll's glands—ordinary sweat glands
▪ Upper and lower eyelids completely cover the closed eye

CONJUNCTIVAE

▪ Thin mucous membranes that line the inner surface of each eyelid and the anterior portion of the sclera
▪ Guard the eye from invasion by foreign matter

LACRIMAL APPARATUS

▪ Structures include the lacrimal glands, punctum, lacrimal sac, and nasolacrimal duct
▪ Lubricate and protect the cornea and conjunctivae
▪ Produce and absorb tears
▪ Tears keep the cornea and conjunctivae moist and protect against bacterial invasion

INTRAOCULAR EYE STRUCTURES

▪ Directly involved with vision
▪ Consists of an anterior segment (which includes the sclera, cornea, iris, pupil, anterior chamber, aqueous humor, lens, ciliary body, and posterior chamber) and a posterior segment (which includes the vitreous humor, posterior sclera, choroid, and retina)

SCLERA AND CORNEA

▪ Sclera—maintains size and form of eyeball
▪ Cornea—smooth, transparent tissue that is highly sensitive to touch and is kept moist by tears

IRIS AND PUPIL

- Iris—circular contractile disk with opening for the pupil
- Pupil size controlled by involuntary dilatory and sphincter muscles that regulate light entry

ANTERIOR CHAMBER AND AQUEOUS HUMOR

- Anterior chamber—a cavity bounded in front by the cornea and behind by the lens and iris; filled with aqueous humor

LENS

- Situated directly behind the iris at the pupillary opening
- Refracts and focuses light onto the retina

CILIARY BODY

- Controls lens thickness
- Regulates light focused through the lens onto the retina

POSTERIOR CHAMBER

- Space directly posterior to the iris but anterior to the lens
- Filled with aqueous humor

VITREOUS HUMOR

- Thick, gelatinous material that fills the space behind the lens
- Maintains placement of the retina and spherical shape of the eyeball

POSTERIOR SCLERA AND CHOROID

- Posterior sclera is a white, opaque, fibrous layer—covers the posterior segment of the eyeball
- Choroid lies beneath the posterior sclera

RETINA

- Innermost coat of the eyeball
- Receives visual stimuli and sends them to the brain

OPTIC DISK

- Well-defined, 1.5-mm round or oval area on the retina
- Allows the optic nerve to enter the retina

PHYSIOLOGIC CUP

- Light-colored depression within the optic disk

PHOTORECEPTOR NEURONS

- Rods and cones—responsible for vision

MACULA

- Main receptor for vision and color

Ears

- Organs of hearing
- Maintain the body's equilibrium

EXTERNAL EAR STRUCTURES

- Consists of the auricle (pinna) and external auditory canal

MIDDLE EAR STRUCTURES

- Also called *tympanic cavity*
- Air-filled cavity within the temporal bone

TYMPANIC MEMBRANE

- Transmits sound vibrations to the internal ear

EUSTACHIAN TUBE

- Extends from the middle ear cavity to the nasopharynx
- Equalizes pressure against the tympanic membrane and prevents rupture

OVAL WINDOW

- Fenestra ovalis—opening in the wall between the middle and inner ears
- Transmits vibrations to the inner ear

ROUND WINDOW

- Enclosed by the secondary tympanic membrane
- Transmits vibrations to the inner ear

SMALL BONES

- Conduct vibratory motion of the tympanum to oval window
  – Malleus (hammer)—attaches to the tympanic membrane and transfers sound to the incus
  – Incus (anvil)—articulates the malleus and the stapes and carries vibration to the stapes
  – Stapes (stirrup)—connects vibratory motion from the incus to the oval window

INNER EAR STRUCTURES

- Vibration excites receptor nerve endings
- Composed of bony and membranous labyrinth
- Inner ear contains the vestibule, cochlea, and semicircular canals

VESTIBULE

- Serves as the entrance to the inner ear
- Senses gravity changes and linear and angular acceleration

COCHLEA

- Bony, spiraling cone
- Contains the cochlear duct—houses the organ of Corti (receptor organ that transmits sound to the acoustic nerve)

SEMICIRCULAR CANALS

- Three semicircular canals project from the posterior aspect of the vestibule
- Each oriented in superior, posterior, or lateral plane
- Crista ampullaris—contains hair cells that are stimulated by sudden movements or changes in the rate or direction of movement

HEARING PATHWAYS

- Sound waves travel through the ear by air conduction and bone conduction
- Vibrations transmitted through air and bone stimulate nerve impulses in the inner ear
- Cochlear branch of the acoustic nerve transmits these vibrations to the auditory area of the cerebral cortex
- Cerebral cortex then interprets the sound

Nose and mouth

- Nose is the sense organ for smell
- Olfactory (smell) receptors, consist of hair cells—highly sensitive but easily fatigued
- Tongue and the roof of the mouth contain most of the receptors for the taste nerve fibers (located in branches of cranial nerves VII and IX)
- Taste buds are stimulated by chemicals and respond to four taste sensations: sweet, sour, bitter, and salty

■ Cross-training

Test your knowledge of terms related to the nervous system by completing the following crossword puzzle.

Across

2. Fibers that conduct impulses away from the cell body

3. Basic unit of the nervous system

7. Brain lobe influencing personality and judgment

8. White, fatty material covering the axon

11. The supportive cells of the nervous system

12. Portion of the brain stem responsible for pupillary reflexes

13. Fibers that conduct impulses toward the cell body

14. Phagocytic cells that ingest and digest microorganisms and waste products from injured neurons

Down

1. The largest part of the brain

2. Nerve cells that supply nutrients to neurons and help them maintain their electrical potential

3. The conduction of electrochemical impulses throughout the nervous system

4. Brain region that helps maintain muscle tone, coordinate muscle movement, and control balance

5. Portion of the brain responsible for controlling body temperature

6. The artery supplying blood to the posterior brain

10. Portion of the brain stem controlling respirations

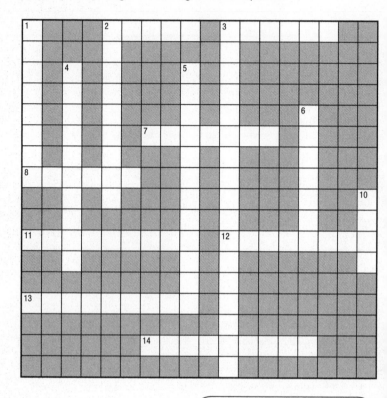

Thermal stimuli such as heat and cold can trigger the conduction of electrochemical impulses through the nervous system. Given a choice, I'll take heat over cold any day!

Boxing match

Use the syllables in the boxes to create the names of the four types of neuroglia using the clues provided. The number of syllables for each answer is shown in parentheses. Use each syllable only once.

AS	CROG	DEN	DROG	DY	EPEN	GO	I
LIA	LIA	LIA	MAL	MI	OL	TROG	

1. Supply nutrients to neurons   (3) __ __ __ __ __ __ __ __

2. Cells that help produce cerebrospinal fluid   (3) __ __ __ __ __ __ __ __

3. Ingest microorganisms and wastes   (3) __ __ __ __ __ __ __ __

4. Support and insulate axons   (6) __ __ __ __ __ __ __ __ __ __ __ __ __ __ __ __

Hit or miss

Some of the following statements are true; the others are false. Mark each accordingly.

_____ 1. Astroglia exist throughout the nervous system.

_____ 2. Neuron activity can only be provoked through chemical stimuli.

_____ 3. Dendrites conduct impulses away from the cell body.

_____ 4. Ependymal cells line the ventricles of the brain and the choroid plexuses and help produce cerebrospinal fluid.

_____ 5. The nodes of Ranvier produce the myelin sheath.

_____ 6. Typical neurons have one axon and multiple dendrites.

_____ 7. Astroglia form part of the blood-brain barrier.

_____ 8. Oligodendroglia form myelin sheaths.

_____ 9. The myelin sheath helps increase conduction of neurotransmitters.

_____ 10. The central nervous system consists of the brain and the spinal cord.

The term *glial* is derived from the Greek word for glue. Glial cells "glue" neurons together.

■ Finish line

This illustration shows a typical neuron. Label the parts of the neuron where indicated.

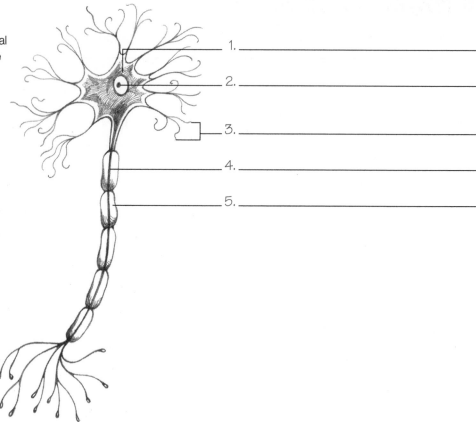

1. _____

2. _____

3. _____

4. _____

5. _____

■ Match point

Match the brain structures on the left with their functions, listed on the right.

1. Cerebrum _____

2. Cerebral cortex _____

3. Grey matter _____

4. White matter _____

5. Basal ganglia _____

6. Corpus callosum _____

7. Cerebellum _____

8. Brain stem _____

9. Diencephalon _____

10. Medulla oblongata _____

11. Limbic system _____

A. Mass of nerve fibers that allows communication between the right and left brain hemispheres

B. Portion of brain stem influencing cardiac, respiratory, and vasomotor functions

C. Myelinated nerve fibers found within the cerebrum

D. Lies between the cerebrum and midbrain; consists of the thalamus and hypo-thalamus

E. Brain region whose functions include maintaining muscle tone, coordinating muscle movement, and controlling balance

F. Houses nerve center that controls sensory and motor activities and intelligence

G. Brain region consisting of the midbrain, pons, and medulla oblongata

H. Outer layer of cerebrum

I. Primitive brain area deep in temporal lobe that initiates basic drives

J. Unmyelinated nerve fibers found in the outer layer of the cerebral cortex

K. Found in white matter; control motor coordination and steadiness

Power stretch

Unscramble the words on the left to reveal the four lobes of the brain. Then draw a line from the boxes on the left to the list on the right, linking each lobe to its particular functions.

FATLORN

_ _ _ _ _ _ _

APEARLIT

_ _ _ _ _ _ _ _

ORALTEMP

_ _ _ _ _ _ _ _

CALICOTIP

_ _ _ _ _ _ _ _ _

A. Influences personality, judgment, abstract reasoning, and social behavior

B. Controls storage and recall of memories

C. Interprets size, shape, distance, and texture

D. Influences language expression

E. Interprets and integrates sensations, including pain, temperature, and touch

F. Important for awareness of body shape

G. Controls hearing and language comprehension

H. Functions mainly to interpret visual stimuli

I. Influences voluntary movement (in the motor portion)

Finish line

Label each of the major brain structures in the illustration below.

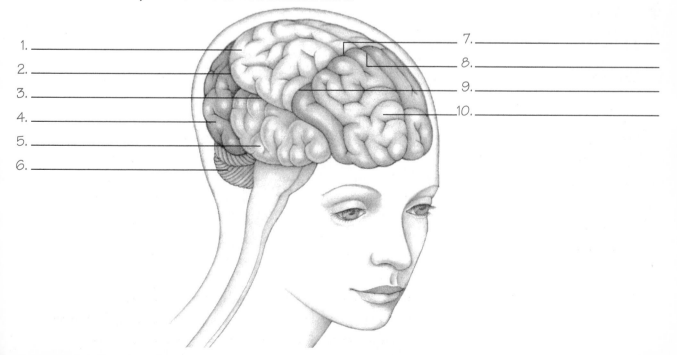

1. _____

2. _____

3. _____

4. _____

5. _____

6. _____

7. _____

8. _____

9. _____

10. _____

Circuit training

The brain receives oxygenated blood through four major arteries. The squares below list various aspects of blood's path to the brain. Draw arrows between the squares to establish the correct circulatory route.

...two internal carotids, which...

Subclavian arteries branch to form...

...supplies blood to the posterior brain.

...two vertebral arteries, which...

Common carotids branch to form...

...supply blood to the anterior and middle brain.

...become the basilar artery, which...

Train your brain

Sound out each group of pictures or symbols to discover a key component of the brains circulatory system.

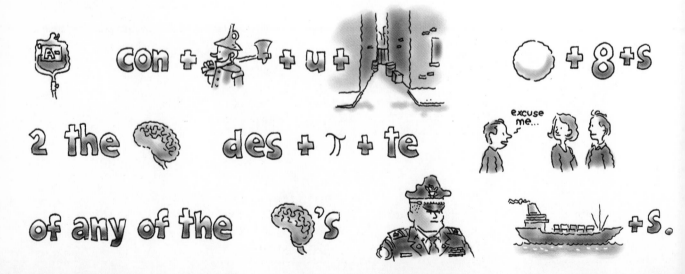

Match point

Match the terms on the left with their definitions on the right.

1. Ganglia _____
2. Efferent neural pathways _____
3. Afferent neural pathways _____
4. Dura mater _____
5. Endosteal dura _____
6. Meningeal dura _____
7. Arachnoid membrane _____
8. Subdural space _____
9. Subarachnoid space _____
10. Pia mater _____

A. Tough, fibrous, leatherlike tissue covering the brain

B. Thin, fibrous membrane that hugs the brain and spinal cord

C. Motor, or descending, pathways

D. Continuous, delicate layer of connective tissue that covers and contours the spinal tissue and brain

E. Forms the periosteum of the skull and is continuous with the lining of the vertebral canal

F. Lies between the dura mater and the arachnoid membrane

G. Knotlike masses of nerve cell bodies on the dorsal roots of spinal nerves

H. Thick membrane that covers the brain, dipping between the brain tissue and providing support and protection

I. Lies between the arachnoid membrane and the pia mater

J. Sensory, or ascending pathways

Circuit training

Sensory and motor impulses travel along different neural pathways. Draw arrows between the boxes, linking them to form the correct neural pathways.

Pain and temperature sensation	Impulses enter spinal cord through ganglia	Impulses travel along efferent neural pathways	
			Impulses enter the thalamus
Touch, pressure, and vibration sensation	Impulses enter spinal cord through dorsal horn	Impulses immediately cross to opposite side of cord and travel up the spinothalamic tract	
			Impulses reach the muscles
Motor impulses	Impulses originate in motor cortex of frontal lobe	Impulses travel up cord in dorsal column to the medulla, where they cross over	

■ Power stretch

Upper motor neurons transmit impulses to the lower motor neurons using two different systems. Unscramble the words in the boxes on the left to reveal the names of these two systems. Then draw a line from the boxes on the left to the list on the right, linking each system to its particular characteristics.

A. Controls gross motor movements

B. Responsible for fine, skilled movements of skeletal muscle

C. Impulses travel from the motor cortex through the internal capsule to the medulla

D. Impulses originate in the premotor area of the frontal lobes and travel to the pons

E. Impulses cross to the opposite side at the pons and then travel down the spinal cord to the anterior horn

F. Impulses cross to the opposite side at the medulla and continue down the spinal cord

G. Impulses are relayed to the lower motor neurons at the anterior horn; these neurons then carry the impulses to the muscles

AMIDLAPRY

_ _ _ _ _ _ _ _ _

ADEARLYPARTMIX

_ _ _ _ _ _ _ _ _ _ _ _ _ _

■ Train your brain

Sound out each group of pictures or symbols to discover information about the spinal cord.

1.

2.

 bod + ee in   + ral are

 to voluntary re + + ivity.

■■
■ Parallel bars

Complete each of the analogies below by determining the relationship between the given word pair and then determining the missing term in the second word pair.

1. Biceps : _____ :: Babinski's : Superficial
2. Pyramidal : Skilled :: _____ : Gross
3. Efferent : Motor :: _____ : Sensory
4. _____ : Pain :: Hypothalamus : Body temperature
5. Largest : Cerebrum :: Second largest : _____

■■
■ Match point

Match the reflex listed on the left with its action, listed on the right.

1. Biceps reflex _____
2. Triceps reflex _____
3. Brachioradialis reflex _____
4. Patellar reflex _____
5. Achilles reflex _____
6. Babinski's reflex _____
7. Cremasteric reflex _____
8. Abdominal reflexes _____
9. Plantar flexion of the toes _____

A. Upward movement of the great toe and fanning of the little toes in response to stimulation of the outer margin of the sole of the foot

B. Contracts the biceps muscle and forces flexion of the forearm

C. Toes curl downward when the lateral sole of an adult's foot is stroked from heel to great toe with a tongue blade

D. Causes supination of the hand and flexion of the forearm at the elbow

E. Forces plantar flexion of the foot at the ankle

F. Contracts the triceps muscle and forces extension of the forearm

G. Forces contraction of the quadriceps muscle in the thigh with extension of the leg

H. Contraction of the cremaster muscle and elevation of the testicle in response to stroking of the inner thigh

I. Movement of the umbilicus in response to stroking the sides of the abdomen above and below the umbilicus

■ You make the call

There are five deep tendon reflexes. Using the following illustrations as a guide, name each of the reflexes and describe the methods for eliciting them.

1. _____

2. _____

3. _____

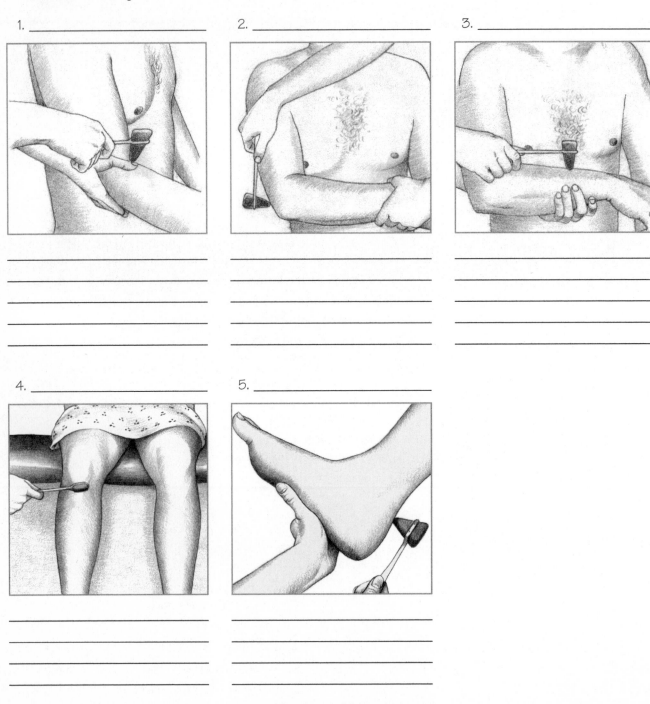

_____

_____

_____

_____

_____

4. _____

5. _____

_____

_____

_____

_____

_____

■ Hit or miss

Some of the following statements are true; the others are false. Mark each accordingly.

_____ 1. The central nervous system consists of the brain and spinal cord.

_____ 2. The peripheral nervous system consists of the cranial nerves, spinal nerves, and autonomic nervous system (ANS).

_____ 3. There are 31 pairs of cranial nerves.

_____ 4. Cranial nerves transmit motor or sensory messages (or both) between the brain or brain stem and the head and neck.

_____ 5. All cranial nerves exit from the midbrain, pons, or medulla oblongata of the brain stem.

_____ 6. Each spinal nerve consists of afferent (sensory) and efferent (motor) neurons.

_____ 7. The ANS innervates all the skeletal muscles.

_____ 8. The ANS carries messages to and from body regions called dermatomes.

> Each spinal nerve gets its name from the vertebra immediately below its exit point from the spinal cord.

■ Power stretch

The ANS has two major subdivisions. Unscramble the words in the boxes on the left to reveal the names of these two nervous systems. Then draw a line from the boxes on the left to the list on the right, linking each system to its physiologic responses.

A. Bronchial smooth-muscle constriction

B. Vasoconstriction

C. Reduction of heart rate, contractility, and conduction velocity

D. Elevated blood pressure

E. Increased heart rate and contractility

F. Increased GI tract tone and peristalsis, with sphincter relaxation

G. Enhanced blood flow to skeletal muscles

H. Increased respiratory rate

I. Vasodilation of external genitalia, causing erection

J. Increased pancreatic, salivary, and lacrimal secretions

K. Smooth-muscle relaxation of the bronchioles, GI tract, and urinary tract

L. Sphincter contraction

M. Pupil constriction

N. Pupillary dilation and ciliary muscle relaxation

O. Increased sweat gland secretion

P. Reduced pancreatic secretion

CAMPSITETHY

_ _ _ _ _ _ _ _ _ _ _ _

APHARMACYTESTPI

_ _ _ _ _ _ _ _ _ _ _ _ _ _ _

■ Finish line

As this illustration reveals, 10 of the 12 pairs of cranial nerves (CNs) exit from the brain stem. The remaining two pairs exit from the forebrain. Using the clues provided about each nerve's function, insert the name and number of each nerve.

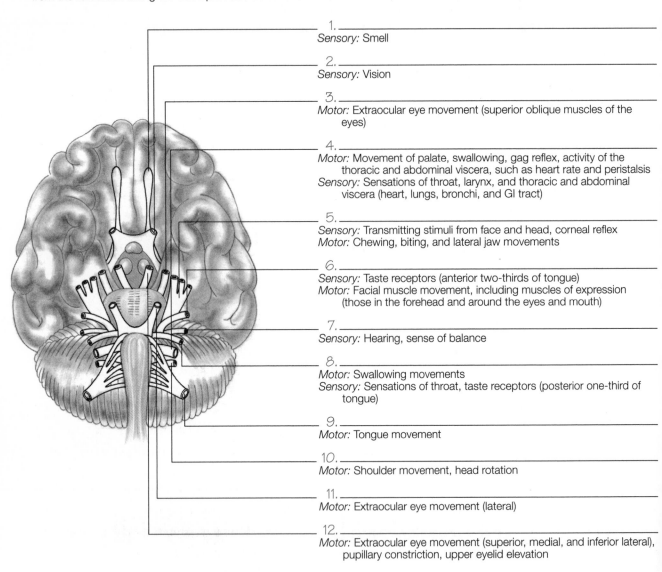

1. _____
Sensory: Smell

2. _____
Sensory: Vision

3. _____
Motor: Extraocular eye movement (superior oblique muscles of the eyes)

4. _____
Motor: Movement of palate, swallowing, gag reflex, activity of the thoracic and abdominal viscera, such as heart rate and peristalsis
Sensory: Sensations of throat, larynx, and thoracic and abdominal viscera (heart, lungs, bronchi, and GI tract)

5. _____
Sensory: Transmitting stimuli from face and head, corneal reflex
Motor: Chewing, biting, and lateral jaw movements

6. _____
Sensory: Taste receptors (anterior two-thirds of tongue)
Motor: Facial muscle movement, including muscles of expression (those in the forehead and around the eyes and mouth)

7. _____
Sensory: Hearing, sense of balance

8. _____
Motor: Swallowing movements
Sensory: Sensations of throat, taste receptors (posterior one-third of tongue)

9. _____
Motor: Tongue movement

10. _____
Motor: Shoulder movement, head rotation

11. _____
Motor: Extraocular eye movement (lateral)

12. _____
Motor: Extraocular eye movement (superior, medial, and inferior lateral), pupillary constriction, upper eyelid elevation

Jumble gym

Use the clues to help you unscramble the names of the structures in the eye. Then use the circled letters to answer the question posed.

The eye is only 1" in diameter, and yet it contains about 70% of the body's sensory receptors.

Question: **What device can be used to assess the inner structures of the eye?**

1. White covering of four-fifths of eyeball

C L E A R S __ __ __ __ __ ◯

2. Smooth, transparent tissue that is continuous with the sclera at the limbus

E A R C O N __ __ __ __ ◯ __

3. Circular contractile disk that contains smooth and radial muscles

I S I R __ __ __ ◯

4. Opening in the center of the iris

L I P U P ◯ __ ◯ __ __

5. Cavity bounded in front by the cornea and behind by the lens and iris

R E A R I T O N   M A R C H B E __ __ __ __ __ __ ◯ __   __ ◯ __ ◯ __ __ __

6. Situated directly behind the iris at the pupillary opening

S N E L ◯ __ __ __

7. Controls lens thickness

R A I L I C Y   B Y D O ◯ __ __ __ __ __ __   __ __ __ __

8. Thick, gelatinous material that fills the space behind the lens

R I V E O U S T   R H O M U __ __ __ __ __ ◯ __ __   __ __ __ ◯ __

9. Receives visual stimuli and sends them to the brain

N E A R I T __ __ ◯ __ __ __

10. Lies beneath the posterior sclera and contains many small arteries and veins

D O C H I O R __ ◯ __ __ __ __ __

Answer: __ __ __ __ __ __ __ __ __ __ __ __ __ __ __

Finish line

Some intraocular structures are visible to the naked eye. Others are visible only with an ophthalmoscope. Test your knowledge of the eye by naming the structures shown in the illustration below.

1. _____
2. _____
3. _____
4. _____
5. _____
6. _____
7. _____
8. _____
9. _____
10. _____
11. _____
12. _____

13. _____
14. _____
15. _____

For hearing to occur, sound waves travel through the ear by two pathways—air conduction and bone conduction.

Match point

Match the ear structures listed on the left with their definitions, listed on the right.

1. Tympanic cavity _____

2. Tympanic membrane _____

3. Eustachian tube _____

4. Oval window (fenestra ovalis) _____

5. Stapes _____

6. Vestibule _____

7. Cochlea _____

8. Organ of Corti _____

9. Saccule and utricle _____

10. Semicircular canals _____

A. Tiny bone of the middle ear that transmits vibrations to the inner ear

B. Air-filled cavity within the hard portion of the temporal bone

C. Projected from the posterior aspect of the vestibule

D. Allows the pressure against inner and outer surfaces of the tympanic membrane to equalize, preventing rupture

E. Membranous sacs in inner ear that sense gravity changes and linear and angular acceleration

F. Layers of skin, fibrous tissue, and a mucus membrane that transmit sound vibrations to the internal ear

G. Bony, spiraling cone extending from the anterior part of the vestibule

H. Opening in the wall between the middle and inner ears

I. Serves as the entrance to the inner ear

J. The receptor organ for hearing

6

Endocrine system

Endocrine system review

Components

※ Glands—specialized cell clusters or organs
※ Hormones—chemical substances secreted by glands in response to stimulation
※ Receptors—protein molecules that trigger specific physiologic changes in a target cell in response to hormonal stimulation

Glands

Pituitary gland

※ Also called the *hypophysis* or *master gland*
※ Rests in the sella turcica
※ Connects with hypothalamus via the infundibulum

ANTERIOR PITUITARY

※ Also called *adenohypophysis*
※ Larger region of the pituitary
※ Produces at least six hormones:
  – Growth hormone (GH), or somatotropin
  – Thyroid-stimulating hormone (TSH), or thyrotropin
  – Corticotropin (ACTH)
  – Follicle-stimulating hormone (FSH)
  – Luteinizing hormone (LH)
  – Prolactin

POSTERIOR PITUITARY

※ Storage area for antidiuretic hormone (ADH), or vasopressin, and oxytocin

Thyroid gland

※ Has two lateral lobes that join with a narrow tissue bridge and give the gland its butterfly shape
※ Lobes function as one unit to produce the hormones triiodothyronine (T_3), thyroxine (T_4), and calcitonin
※ Thyroid hormone:
  – Consists of T_3 and T_4
  – The body's major metabolic hormone
  – Regulates metabolism by speeding cellular respiration
※ Calcitonin
  – Maintains the blood's calcium level

Parathyroid glands

※ Work together as a single gland and produce parathyroid hormone (PTH), which regulates the blood's calcium balance and increases the movement of phosphate ions from blood to urine for excretion

Adrenal glands

※ Each lies on top of a kidney
※ Contain two distinct structures—the adrenal cortex and the adrenal medulla—that function as separate endocrine glands

ADRENAL CORTEX

※ Three zones:
  – Zona glomerulosa (outermost zone)—produces mineralocorticoids, primarily aldosterone
  – Zona fasciculata (middle and largest zone)—produces the glucocorticoids cortisol (hydrocortisone), cortisone, and corticosterone and small amounts of androgen and estrogen
  – Zona reticularis (innermost zone)—produces mainly glucocorticoids and some sex hormones

ADRENAL MEDULLA

※ Produces two catecholamines

Pancreas

※ Performs endocrine and exocrine functions
※ Made up primarily of acinar cells, which regulate pancreatic exocrine function
※ Endocrine cells contain alpha, beta, and delta cells that produce important hormones
  – Alpha cells produce glucagon
  – Beta cells produce insulin
  – Delta cells produce somatostatin

Thymus

※ Produces T cells important in cell-mediated immunity
※ Also produces the peptide hormones thymosin and thymopoietin, which promote growth of peripheral lymphoid tissue

Pineal gland

※ Produces melatonin

Gonads

OVARIES

- Produce ova and the steroidal hormones estrogen and progesterone
- Four functions:
  - Promote development and maintenance of female sex characteristics
  - Regulate the menstrual cycle
  - Maintain the uterus for pregnancy
  - Prepare the mammary glands for lactation (along with other hormones)

TESTES

- Produce spermatozoa and the male sex hormone testosterone
- Testosterone stimulates and maintains male sex characteristics

Hormones

- Complex chemical substances that trigger or regulate the activity of an organ or a group of cells
- Classified by molecular structure as polypeptides, steroids, or amines

Polypeptides

- Protein compounds that include:
  - Anterior pituitary hormones—GH, TSH, FSH, LH, and prolactin
  - Posterior pituitary hormones—ADH and oxytocin
  - PTH
  - Pancreatic hormones—insulin and glucagon

Steroids

- Derived from cholesterol and include:
  - Adrenocortical hormones—secreted by the adrenal cortex (aldosterone and cortisol)
  - Sex hormones—secreted by the gonads (estrogen and progesterone in females and testosterone in males)

Amines

- Derived from tyrosine and include:
  - Thyroid hormones (T_4 and T_3)
  - Catecholamines (epinephrine, norepinephrine, and dopamine)

Hormone release and transport

- ACTH and cortisol are released in spurts in response to body rhythm cycles
- Secretion of PTH and prolactin occurs fairly evenly throughout the day
- Secretion of insulin by the pancreas has both steady and sporadic release patterns

Hormonal action

- When a hormone reaches its target site, it binds to a specific receptor on the cell membrane or within the cell
- After binding, each hormone produces unique physiologic changes

Hormonal regulation

- Feedback mechanism regulates hormone production and secretion
- Released hormone travels to target cells, where a receptor molecule recognizes it and binds to it

Mechanisms that control hormone release

PITUITARY-TARGET GLAND AXIS

- Pituitary gland regulates other endocrine glands and their hormones, including:
  - Corticotropin—regulates adrenocortical hormones
  - TSH—regulates T_4 and T_3
  - LH—regulates gonadal hormones
- Pituitary gland gets feedback about target glands by continuously monitoring levels of hormones produced by these glands
- If a change occurs, the pituitary gland corrects it

HYPOTHALAMIC-PITUITARY-TARGET GLAND AXIS

- Hypothalamus also produces trophic hormones that regulate anterior pituitary hormones
- By controlling anterior pituitary hormones, which regulate the target gland hormones, the hypothalamus affects target glands as well

CHEMICAL REGULATION

- Endocrine glands not controlled by the pituitary gland may be controlled by specific substances that trigger gland secretions

NERVOUS SYSTEM REGULATION

- Hypothalamus controls pituitary hormones
- Nervous system stimuli affect ADH levels
- Autonomic nervous system controls catecholamine secretion

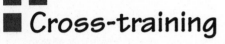

Cross-training

Test your knowledge of terms related to the endocrine system by completing this crossword puzzle.

Across

1. Type of cell that makes up most of the pancreas
4. Hormone thought to regulate circadian rhythms
7. Protein compounds made of many amino acids that are connected by peptide bonds
10. Include the ovaries in females and testes in males
12. An essential amino acid found in most proteins
13. Complex chemical substances that trigger or regulate the activity of an organ or a group of cells

Down

2. Another name for the anterior pituitary gland
3. Substance from which steroids are derived
5. Another name for antidiuretic hormone
6. Gland located below the sternum that contains lymphatic tissue
7. Body's smallest known endocrine glands
8. Narrow tissue bridge connecting the two lobes of the thyroid gland
9. Another name for the pituitary gland
11. Glands that lie on top of each kidney

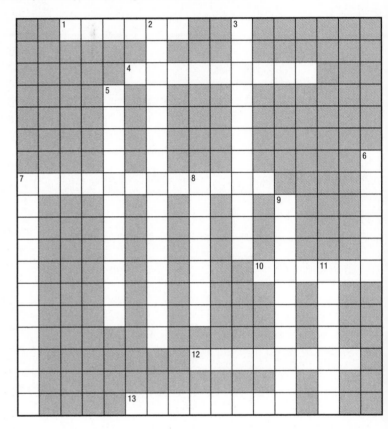

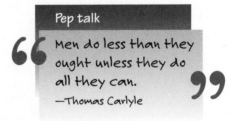

Pep talk

Men do less than they ought unless they do all they can.
—Thomas Carlyle

Finish line

This illustration shows the major structures of the endocrine system. Label all of the components.

1. _____

2. _____

3. _____

4. _____

5. _____

6. _____

The three major components of the endocrine system are glands, hormones, and receptors. All three work together to regulate the body's metabolic activity.

Also known as...

Hormones commonly go by more than one name. Look at each of the pairs below and circle the pairs that correctly match two names for the same hormone.

1. Growth hormone: Somatotropin

2. Thyroid stimulating hormone: Corticotropin

3. Luteinizing hormone: Prolactin

4. Antidiuretic hormone: Vasopressin

5. Triiodothyronine: T_4

■ Match point

Match the glands and hormones listed on the left with their characteristics or functions, listed on the right.

1. Anterior pituitary _____
2. Posterior pituitary _____
3. Thyroid _____
4. Hypothalamus _____
5. Thyroid hormone _____
6. Calcitonin _____
7. Parathyroid glands _____
8. Parathyroid hormone _____
9. Pancreas _____
10. Adrenal medulla _____
11. Thymus _____
12. Ovaries _____
13. Testes _____

A. Maintains the calcium level of blood
B. Helps regulate the blood's calcium balance
C. Produce spermatozoa and testosterone
D. Helps control some endocrine glands by neural and hormonal stimulation
E. Produces T cells as well as the peptide hormones thymosin and thymopoietin
F. The largest of the two regions in the pituitary gland
G. Embedded on the posterior surface of the thyroid gland
H. Produce ova as well as the hormones estrogen and progesterone
I. Serves as a storage area for antidiuretic hormone and oxytocin
J. Regulates metabolism by speeding cellular respiration
K. Performs both endocrine and exocrine functions
L. Consists of two lateral lobes, one on each side of the trachea
M. Produces epinephrine and norepinephrine

■ Power stretch

The two adrenal glands each lie on top of a kidney. These almond-shaped glands contain two distinct structures—the adrenal cortex and the adrenal medulla—that function as separate endocrine glands. Unscramble the words on the left to reveal the three zones, or cell layers, of the adrenal cortex. Then draw a line from each box to the specific characteristics of each zone, listed on the right.

AZON OAGLESOLRUM

_ _ _ _

_ _ _ _ _ _ _ _ _ _ _

NAZO ACACTUSFAIL

_ _ _ _

_ _ _ _ _ _ _ _ _ _ _

NOZA CURLERSIAIT

_ _ _ _

_ _ _ _ _ _ _ _ _ _ _

A. Outermost zone
B. Helps regulate metabolism and resistance to stress
C. Produces mainly glucocorticoids and some sex hormones
D. Helps maintain fluid balance by increasing sodium reabsorption
E. Produces mineralcorticoids (primarily aldosterone)
F. Produces hydrocortisone, cortisone, and corticosterone
G. Middle and largest zone
H. Innermost zone
I. Produces small amounts of sex hormones androgen and estrogen

■ Hit or miss

Some of the following statements are true; the others are false.
Mark each accordingly.

Mechanisms that control hormone release

_____ 1. The islets of Langerhans cells make up most of the pancreas.

_____ 2. The endocrine cells of the pancreas are called the _islet cells_.

_____ 3. The islets contain alpha, beta, and delta cells.

_____ 4. Beta cells produce glucagon.

_____ 5. Glucagon raises blood glucose levels by triggering the breakdown of glycogen to glucose.

_____ 6. Insulin lowers blood glucose levels by stimulating the release of growth hormone.

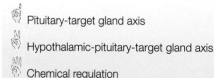

Pituitary-target gland axis

Hypothalamic-pituitary-target gland axis

Chemical regulation

Nervous system regulation

_____ 7. Following endocrine gland stimulation, all hormones are released in a similar pattern.

_____ 8. When a hormone reaches its target site, it binds to a specific receptor on the cell membrane or within the cell.

_____ 9. A hormone acts only on a cell that has a receptor specific to that hormone.

_____ 10. The effects of a particular hormone are the same, regardless of the target site.

■ Power stretch

Hormones are classified by their molecular structure. Unscramble the words on the left to reveal the classes of hormones. Then draw lines to link each class to its specific hormones, listed on the right.

TIDYSLEEPPOP

— — — — — — — — — — — — — —

DIRTOESS

— — — — — — — —

INSEAM

— — — — — —

A. Posterior pituitary hormones

B. Thyroid hormones

C. Anterior pituitary hormones

D. Adrenocortical hormones (secreted by adrenal cortex)

E. Catecholamines

F. Parathyroid hormones

G. Sex hormones (secreted by gonads)

H. Pancreatic hormones

■ You make the call

The negative feedback mechanism that helps regulate the endocrine system may be simple or complex. Using the illustration below as a guide, describe the processes of simple and complex feedback.

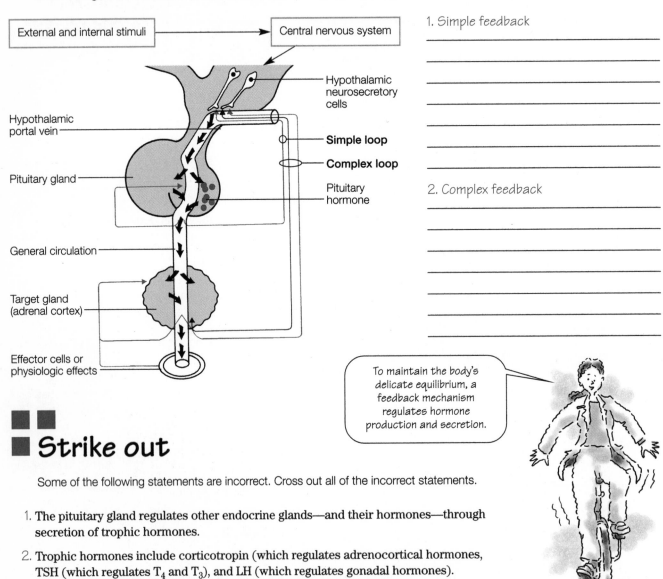

1. Simple feedback

_____

_____

_____

_____

_____

_____

_____

2. Complex feedback

_____

_____

_____

_____

_____

_____

> To maintain the body's delicate equilibrium, a feedback mechanism regulates hormone production and secretion.

■ Strike out

Some of the following statements are incorrect. Cross out all of the incorrect statements.

1. The pituitary gland regulates other endocrine glands—and their hormones—through secretion of trophic hormones.

2. Trophic hormones include corticotropin (which regulates adrenocortical hormones, TSH (which regulates T_4 and T_3), and LH (which regulates gonadal hormones).

3. By increasing trophic hormones, the pituitary gland decreases target gland hormone levels.

4. The hypothalamus also produces trophic hormones that regulate anterior pituitary hormones.

5. The ability of blood glucose levels to regulate glucagon and insulin release is an example of the hypothalamic-pituitary-target gland axis.

6. Examples of nervous system regulation of hormone release include the stimulation of the posterior pituitary to secrete ADH and oxytocin.

7

Cardiovascular system

Warm-up

Cardiovascular system review

Cardiovascular system basics

- Also called the *circulatory system*
- Consists of the heart, blood vessels, and lymphatics
- Brings oxygen and nutrients to the body's cells
- Removes metabolic waste products
- Carries hormones throughout the body

Components

Heart

- Size of a closed fist
- Lies beneath the sternum in the mediastinum, between the second and sixth ribs
- Rests obliquely, with its right side below and almost in front of the left
- Apex is the point of maximal impulse, where heart sounds are loudest
- Four chambers (two atria and two ventricles)
- Four valves (two atrioventricular [AV] and two semilunar valves)
- Right side pumps blood to the lungs
- Left side pumps blood to the rest of the body
- Surrounded by a sac called the *pericardium*
- Three-layers of wall:
  - Epicardium—outer layer
  - Myocardium—middle layer
  - Endocardium—inner layer

PERICARDIUM

- Fibroserous sac that surrounds the heart and the roots of the great vessels
- Consists of the fibrous pericardium and the serous pericardium
- Pericardial space—area between the fibrous and serous pericardium
- Pericardial fluid lubricates the surfaces of the space and allows the heart to move easily during contraction

FIBROUS PERICARDIUM

- Composed of tough, white fibrous tissue
- Protects the heart

SEROUS PERICARDIUM

- Thin, smooth inner portion

- Two layers:
  - Parietal layer lines the inside of the fibrous pericardium
  - Visceral layer adheres to the surface of the heart

Heart chambers

ATRIA

- Two upper chambers
- Separated by interatrial septum
- Receive blood returning to the heart
- Pump blood to the ventricles
- Right atrium—receives blood from the superior and inferior venae cavae
- Left atrium— receives blood from the two pulmonary veins; smaller but has thicker walls than the right atrium

VENTRICLES

- Two lower chambers
- Separated by the interventricular septum
- Receive blood from the atria
- Right ventricle—pumps blood to the lungs
- Left ventricle—pumps blood through all other vessels of the body; larger than right ventricle

Heart valves

- Two AV valves and two semilunar valves
- Allow forward flow of blood through the heart and prevent backward flow
- Open and close in response to pressure changes caused by ventricular contraction and blood ejection
- AV valves separate the atria from the ventricles
  - Pulmonic valve—prevents backflow from the pulmonary artery into the right ventricle
  - Aortic valve—prevents backflow from the aorta into the left ventricle
- Right AV valve (tricuspid) has three triangular cusps
- Left AV valve (mitral or bicuspid) has two cusps
- Chordae tendineae attach the cusps of the AV valves to papillary muscles in the ventricles

The heart's contraction

- Contraction of the heart causes the blood to move throughout the body
- The conduction system contains pacemaker cells, which have three unique characteristics:
  - Automaticity—ability to generate an electrical impulse automatically

– Conductivity—ability to pass the impulse to the next cell
– Contractility—ability to shorten the fibers in the heart when receiving the impulse

Sinoatrial (SA) node

- Normal pacemaker of the heart
- Generates an impulse between 60 and 100 times per minute
- SA node's firing spreads an impulse throughout the right and left atria, resulting in atrial contraction

Atrioventricular (AV) node

- Situated low in the septal wall of the right atrium
- Slows impulse conduction between the atria and ventricles

Typical impulse conduction

- Generates in SA node → AV node → bundle of His → Purkinje fibers → spreads and tells the blood-filled ventricles to contract

Cardiac cycle

- Period from the beginning of one heartbeat to the beginning of the next
- Two phases: systole and diastole

SYSTOLE

- Ventricles contract
- Blood pressure in the ventricles increases
- AV valves close
- Ventricular blood pressure builds until it exceeds the pressure in the pulmonary artery and the aorta
- Semilunar valves open, and the ventricles eject blood into the aorta and the pulmonary artery
- When the ventricles empty and relax, ventricular pressure falls below that in the pulmonary artery and the aorta

DIASTOLE

- Semilunar valves close to prevent backflow of blood into the ventricles
- AV valves open, allowing blood to flow into the ventricles from the atria
- Ventricles fill and the atria contract to send the remaining blood to the ventricles
- New cardiac cycle begins as the heart enters systole again

Cardiac output

- Amount of blood the heart pumps in 1 minute

Stroke volume

- Amount of blood ejected with each heartbeat
- Depends on contractility, preload, and afterload

Blood flow

- Blood travels through five distinct types of blood vessels: arteries, arterioles, capillaries, venules, veins

Vessel structure

- Differs according to its function and the pressure exerted by the volume of blood

ARTERIES

- Have thick, muscular walls to accommodate the flow of blood at high speeds and pressures

ARTERIOLES

- Have thinner walls than arteries
- Constrict or dilate to control blood flow to the capillaries

VENULES

- Gather blood from the capillaries
- Have thinner walls than arterioles

VEINS

- Have thinner walls than arteries
- Have larger diameters because of the low blood pressures of venous return to the heart

Circulation

PULMONARY CIRCULATION

- Blood travels to the lungs to pick up oxygen and release carbon dioxide
- Unoxygenated blood travels from the right ventricle through the pulmonic valve into the pulmonary arteries
- Blood passes through progressively smaller arteries and arterioles into the capillaries of the lungs
- Blood reaches the alveoli and exchanges carbon dioxide for oxygen
- Oxygenated blood then returns via venules and veins to the pulmonary veins, which carry it back to the heart's left atrium

SYSTEMIC CIRCULATION

- Blood pumped from the left ventricle carries oxygen and other nutrients to body cells and transports waste products for excretion
- The aorta branches into vessels that supply specific organs and areas of the body
- The common carotid artery supplies blood to the brain
- The subclavian artery supplies the arms
- The innominate artery supplies the upper chest
- The aorta descends through the thorax and abdomen and branches supply the organs of the GI and genitourinary systems, spinal column, and lower chest and abdominal muscles
- The aorta divides into the iliac arteries, which further divide into femoral arteries
- Valves in the veins prevent blood backflow
- The veins merge until they form two main branches, the superior vena cava and inferior vena cava, that return blood to the right atrium

CORONARY CIRCULATION

- During left ventricular systole, blood is ejected into the aorta
- During diastole, blood flows out of the heart and then through the coronary arteries to nourish the heart muscle

■ Finish line

This illustration shows the structures of the heart. See if you can label the key structures.

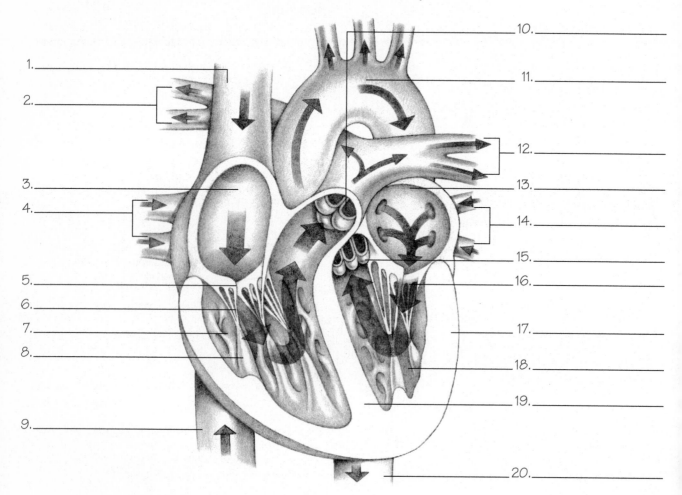

1. _____

2. _____

3. _____

4. _____

5. _____

6. _____

7. _____

8. _____

9. _____

10. _____

11. _____

12. _____

13. _____

14. _____

15. _____

16. _____

17. _____

18. _____

19. _____

20. _____

Match point

Match each of the terms on the left with its definition on the right.

1. Pericardium _____
2. Fibrous pericardium _____
3. Serous pericardium _____
4. Pericardial space _____
5. Epicardium _____
6. Myocardium _____
7. Endocardium _____
8. Chordae tendineae _____
9. Pericardial fluid _____

A. Outer layer of the wall of the heart
B. Fibroserous sac that surrounds the heart and the roots of the great vessels
C. Attach the cusps of the AV valves to papillary muscles in the ventricles
D. Contains striated muscle fibers that cause the heart to contract
E. Tough, white fibrous tissue that fits loosely around the heart, protecting it
F. Space between the fibrous and serous pericardium
G. Heart's inner layer
H. Thin, smooth inner portion of the pericardium
I. Lubricates the surfaces of the pericardial space and allows the heart to move easily during contraction

Circuit training

Trace the circulation of blood on the right side of the heart by linking the appropriate boxes below with red arrows. Trace the circulation of blood on the left side of the heart by linking the appropriate boxes with black arrows.

| Superior vena cava | Pulmonary veins | Lungs | Aorta |

| Tricuspid valve | Mitral valve | Pulmonic valve | Aortic valve |

| Right atrium | Left atrium | Right ventricle | Left ventricle |

■ Cross-training

Test your knowledge of terminology associated with the cardiovascular system by completing this crossword puzzle.

Across

5. Pointed end of the heart

7. Major artery

9. The inherent ability of the myocardium to contract normally

10. Fibroserous sac surrounding the heart and the roots of the great vessels

12. Heart's inner layer

13. Upper chambers of the heart

14. Valve between the right atria and ventricle

Down

1. Heart's outer layer

2. Heart's middle layer

3. Phase of the cardiac cycle when the ventricles relax

4. Cavity between the lungs

6. Heart's two lower chambers

8. Phase of the cardiac cycle when the ventricles contract

9. The body's smallest vessels

11. Valve between the left atria and ventricle

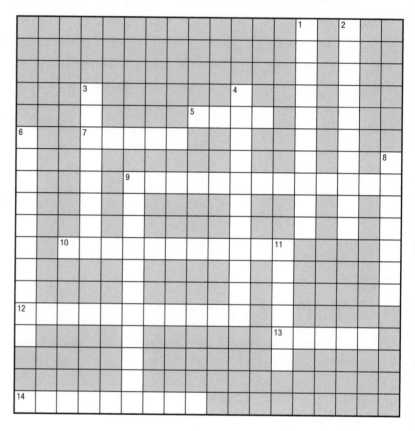

Everyone likes to relax while they eat. Likewise, it's during diastole—or ventricular relaxation—that blood flows through the coronary arteries to nourish the heart muscle.

Strike out

Some of the statements below about the heart are incorrect. Cross out all of the incorrect statements.

1. The heart contains four valves, two AV valves and two semilunar valves.

2. The AV valves separate the aorta from the ventricles.

3. The right AV valve is called the *mitral valve.*

4. The right AV valve prevents backflow from the right ventricle into the right atrium.

5. The left AV valve prevents backflow from the aorta into the left ventricle.

6. One of the semilunar valves is the pulmonic valve.

7. The pulmonic valve prevents backflow from the pulmonary veins into the right atrium.

8. The other semilunar valve is the aortic valve.

9. The aortic valve prevents backflow from the aorta into the left ventricle.

10. Both the tricuspid and mitral valve have three cusps.

11. The semilunar valves have three cusps that are shaped like half-moons.

At rest, only about 20% of my blood goes to skeletal muscles. When I exercise, that percentage can increase to 70%.

Train your brain

Sound out each group of pictures and symbols to reveal information about normal heart function.

■ Parallel bars

Complete each of the analogies below by determining the relationship between the given word pair and then determining the missing term in the second word pair.

1. _____ : Myocardium :: Chamber : Ventricle

2. Fibrous and serous : Pericardium :: _____ and _____ : Serous

3. Inner : _____ :: Outer : Epi

4. Ventricle : Ventricles :: _____ : Atria

5. Left AV : _____ :: Right AV : Tricuspid

6. _____ : Generate :: Conductivity : Pass

7. SA : 60 to 100 :: AV : _____

■ Hit or miss

Some of the following statements about the conduction system are true; the others are false. Mark each accordingly.

_____ 1. The sinoatrial (SA) node is the normal pacemaker of the heart.

_____ 2. The SA node is located deep in the myocardium of the right atrium.

_____ 3. The SA node generates an impulse between 60 and 70 times per minute.

_____ 4. The firing of the SA node spreads an impulse throughout the right and left atria, resulting in atrial contraction.

_____ 5. The AV node is located in the upper third of the left ventricle.

_____ 6. The AV node accelerates impulse conduction between the atria and the ventricle.

_____ 7. From the AV node, the impulse travels to the bundle of His.

_____ 8. From the Bundle of His, the impulse branches to the right and left bundles.

_____ 9. The impulse then travels to the Purkinje fibers and the ventricles contract.

_____ 10. If the SA node fails to fire, the AV node automatically takes over at the same rate.

_____ 11. If the AV node fails to fire, the heart fails to beat.

The conduction system of the heart contains pacemaker cells that have characteristics unique only to them. That makes me feel special!

■ Finish line

This illustration shows the elements of the cardiac conduction system. See if you can name them all.

1. _____

2. _____

3. _____

4. _____

5. _____

6. _____

7. _____

8. _____

Yes, sir, the heart is one amazing electrical appliance. Specialized fibers propagate electrical impulses throughout the heart's cells in a precise sequence thousands of times a day.

Train your brain

Sound out each group of pictures and symbols to reveal information about normal physiology of the heart.

Match point

Match each of the terms on the left with its definition on the right.

1. Systole _____

2. Diastole _____

3. Cardiac output _____

4. Stroke volume _____

5. Preload _____

6. Starling's law _____

7. Contractility _____

8. Afterload _____

9. Resistance _____

A. The amount of blood ejected with each heartbeat

B. The pressure that the ventricular muscle must generate to eject blood from its chamber

C. Phase of the cardiac cycle when ventricles contract

D. Stretching of muscle fibers in the ventricles

E. Pressure ventricle must work against

F. Phase of the cardiac cycle when the ventricles relax

G. The more the heart muscles stretch during diastole, the more forcefully they contract during systole

H. The amount of blood the heart pumps in 1 minute

I. The inherent ability of the myocardium to contract normally

I'll give you a hint: Cardiac output is equal to the heart rate multiplied by the stroke volume.

HRxSV=CO

You make the call

This illustration shows the five events of the cardiac cycle. Using the space provided, explain what's occurring in each stage.

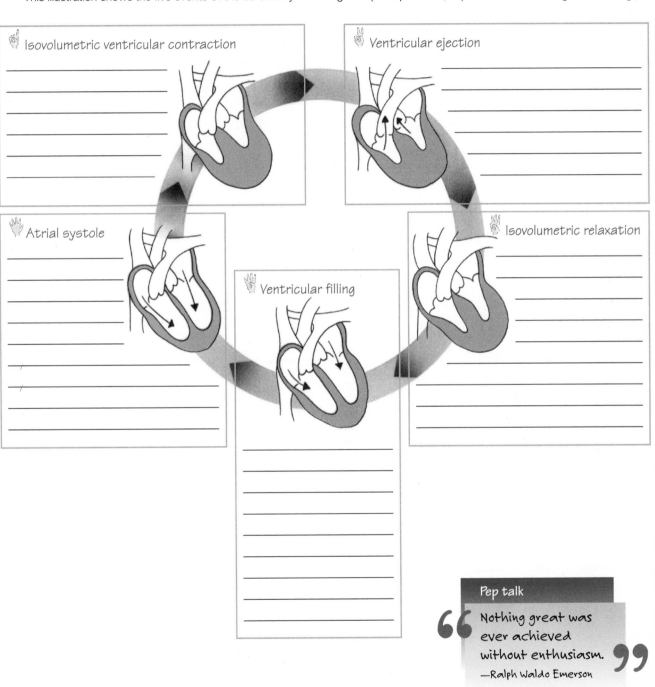

Isovolumetric ventricular contraction

_____
_____
_____
_____
_____
_____

Ventricular ejection

_____
_____
_____
_____
_____

Atrial systole

_____
_____
_____
_____
_____
_____
_____

Ventricular filling

_____
_____
_____
_____
_____
_____
_____
_____

Isovolumetric relaxation

_____
_____
_____
_____
_____
_____
_____

Power stretch

Stroke volume depends on three major factors. Unscramble the jumbled words on the left to reveal the names of these three factors. Then draw lines from each of these boxes to the descriptions on the right that match each term.

PADROLE

_ _ _ _ _ _ _

ACRYLICTINTTO

_ _ _ _ _ _ _ _ _ _ _ _ _

AFLOATRED

_ _ _ _ _ _ _ _ _

A. The balloon's stretch

B. Pressure ventricular muscles must generate to overcome aortic pressure

C. Like blowing up a balloon

D. The balloon's knot

E. Inherent ability of myocardium to contract normally

F. Stretching of muscle fiber in ventricles

Odd man out

In each of the word groupings below, circle the odd man out (the word that doesn't belong) by figuring out the connection between all of the other words in the group.

1. Heart   Lungs   Blood vessels   Lymphatics

2. Pericardium   Epicardium   Endocardium   Myocardium

3. Atria   Mediastinum   Ventricles

4. Bicuspid   Tricuspid   Mitral   Pulmonic

5. Arteries   Venules   Veins   Arterioles   Alveoli   Capillaries

6. Systemic   Intrinsic   Coronary   Pulmonary

Jumble gym

Use the clues to help you unscramble the words related to the vascular system. Then use each of the circled letters to answer the question posed below.

Question: Which blood vessels have walls composed of only a single layer of endothelial cells?

1. Have thick, muscular walls to accommodate the flow of blood at high speeds and pressures

E R A S E R I T ◯ _ _ _ ◯ _ _ _

2. Have thinner walls than arteries; constrict or dilate to control blood flow

R O A S T E R L I E ◯ _ _ _ _ _ _ ◯ _ _

3. Gather blood from capillaries

E L V E S N U _ ◯ _ _ _ _ ◯

4. Have thinner walls than arteries but have larger diameters

V I N E S _ _ ◯ _ _ _

5. Control blood flow into the tissues

C H E S S P R I N T _ ◯ _ ◯ _ ◯ _ _ _ _

6. Prevent the backflow of blood in veins

S L A V E V _ _ ◯ _ _ _

Answer: _ _ _ _ _ _ _ _ _ _ _

Once blood reaches the alveoli, it exchanges carbon dioxide for oxygen.

Thank you, come again!

CO_2

O_2

Circuit training

Trace the path of blood from the heart to the lungs and back again by drawing arrows between the boxes below.

Arteries

Right ventricle

Pulmonary arteries	Arterioles	Lung capillaries

Pulmonary valve

Pulmonary veins	Venules	Alveoli

Left atrium

Veins

■ Finish line

The illustration below shows the body's major arteries and veins. See if you can name them all.

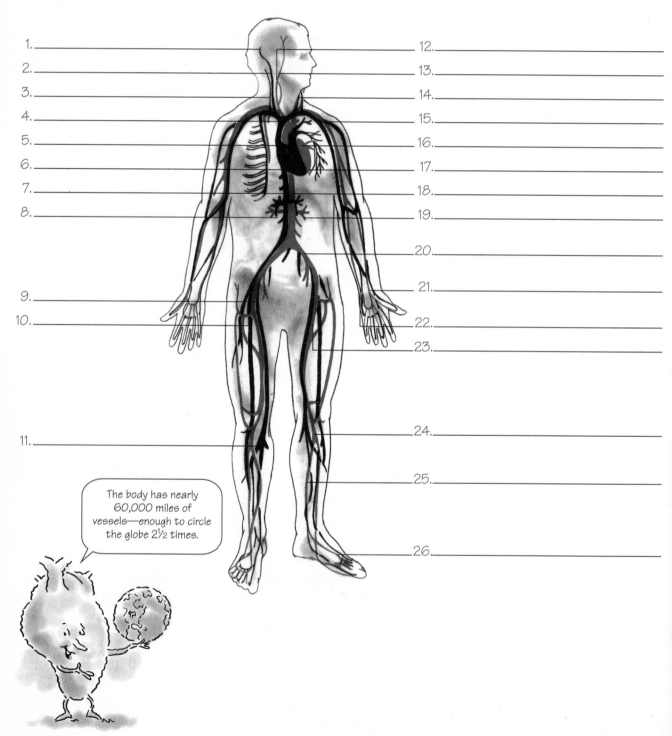

1. _____
2. _____
3. _____
4. _____
5. _____
6. _____
7. _____
8. _____

9. _____
10. _____

11. _____

12. _____
13. _____
14. _____
15. _____
16. _____
17. _____
18. _____
19. _____
20. _____
21. _____
22. _____
23. _____

24. _____

25. _____

26. _____

The body has nearly 60,000 miles of vessels—enough to circle the globe 2½ times.

Train your brain

Sound out each group of pictures and symbols to reveal information about coronary circulation.

Power stretch

Unscramble the words on the left to reveal the names of the major coronary blood vessels. Then draw lines from each of these boxes to the list on the right, linking each vessel to its particular characteristics.

GIRTH CRAYONOR RATRYE

— — — — —
— — — — — — — — —
— — — — — —

FELT RAYCROON TARREY

— — — —
— — — — — — — — —
— — — — — —

ACIDARC VINES

— — — — — — —
— — — — — —

RANCORYO ISNUS

— — — — — — —
— — — — —

A. Supplies blood to the left atrium

B. Supplies blood to the right atrium

C. Lie superficial to the arteries

D. Supplies blood to part of the left atrium

E. Opens into the right atrium

F. Supplies blood to most of the right ventricle

G. Splits into the anterior descending artery and the circumflex artery

H. Largest is the coronary sinus

I. Supplies blood to the inferior part of the left ventricle

J. The largest vein

K. Supplies blood to most of the left ventricle

L. Supplies blood to most of the inter-ventricular septum

Most of the major cardiac veins empty into the coronary sinus, except for the anterior cardiac veins, which empty into the right atrium.

Finish line

The following illustrations show the major vessels involved in coronary circulation. See if you can name them.
Hint: The arteries are tinted red.

Anterior view

1. _____

2. _____

3. _____

4. _____

5. _____

6. _____

7. _____

Posterior view

1. _____

2. _____

3. _____

4. _____

5. _____

Coaching session

Cardiovascular system changes with aging

- Decreased heart size
- Decreased contractile strength
- Decreased resting cardiac output
- Dilation and stretching of veins
- Increased systolic blood pressure
- Left ventricular wall thickening
- Thickened heart valves

8

Hematologic system

Warm-up

Hematologic system review

Hematologic system basics

- Consists of the blood and bone marrow
- Blood delivers oxygen and nutrients to all tissues, removes wastes, and transports gases, blood cells, immune cells, and hormones throughout the body
- Hematologic system manufactures new blood cells through a process called *hematopoiesis*
- Multipotential stem cells in bone marrow give rise to five distinct cell types, called *unipotential stem cells*
- Unipotential cells differentiate into one of the following types of blood cells:
    - Erythrocyte
    - Granulocyte
    - Agranulocyte
    - Platelet

Blood components

- Blood consists of formed elements suspended in fluid called *plasma*
- Formed elements in the blood include:
    - Red blood cells (RBCs), or erythrocytes
    - White blood cells (WBCs), or leukocytes
    - Platelets

Red blood cells

- Transport oxygen and carbon dioxide to and from body tissues
- Contain hemoglobin, the oxygen-carrying substance that gives blood its red color
- Bone marrow releases RBCs into circulation in immature form
- The spleen isolates, old, worn-out RBCs, removing them from circulation

White blood cells

- Five types participate in the body's defense and immune systems
- Classified as granulocytes or agranulocytes

GRANULOCYTES

- Types collectively known as *polymorphonuclear leukocytes*

NEUTROPHILS

- Most numerous granulocytes

- Phagocytic cells that engulf, ingest, and digest foreign materials
- Worn-out neutrophils form the main component of pus
- Bone marrow produces their replacements, called *bands*
- In response to infection, bone marrow must produce many immature cells

EOSINOPHILS

- Account for 0.3% to 7% of circulating WBCs
- involved in the ingestion of antigen-antibody complexes

BASOPHILS

- Usually constitute fewer than 2% of circulating WBCs
- Possess little or no phagocytic ability
- secrete histamine in response to certain inflammatory and immune stimuli
- Histamine makes the blood vessels more permeable and eases the passage of fluids from the capillaries into body tissues

AGRANULOCYTES

- Monocytes and lymphocytes
- Lack specific cytoplasmic granules and have nuclei without lobes

MONOCYTES

- Largest WBCs
- Constitute only 0.6% to 9.6% of WBCs in circulation
- Phagocytic

MACROPHAGES

- May roam freely through the body when stimulated by inflammation
- Defend against infection and dispose of cell breakdown products
- Concentrate in the liver, spleen, and lymph nodes, where they defend against invading organisms
- Ingest microorganisms, cellular debris, and necrotic tissue
- Phagocytize cellular remnants and promote wound healing

LYMPHOCYTES

- Smallest WBCs
- Second most numerous
- Derive from stem cells in the bone marrow
- Two types
    - T lymphocytes directly attack an infected cell
    - B lymphocytes produce antibodies against specific antigens

Platelets

- Fragments split from cells in bone marrow called *megakaryocytes*
- Perform three vital functions:
  1. Initiate contraction of damaged blood vessels to minimize blood loss
  2. Form hemostatic plugs in injured blood vessels
  3. With plasma, provide materials that accelerate blood coagulation

How blood cells clot

- Hemostasis is the complex process by which platelets, plasma, and coagulation factors interact to control bleeding

Extrinsic cascade

- When a blood vessel ruptures, local vasoconstriction and platelet clumping help prevent hemorrhage
- Requires the release of tissue thromboplastin from the damaged cells

Intrinsic cascade system

- Formation of a more stable clot
- Activated by a protein, called *factor XII*
- Final result of the intrinsic cascade is a fibrin clot

Coagulation factors

- Materials that platelets and plasma provide work with coagulation factors to serve as precursor compounds in the clotting (coagulation) of blood

FACTOR I

- Fibrinogen
- High-molecular-weight protein synthesized in the liver
- Converted to fibrin during the coagulation cascade

FACTOR II

- Prothrombin
- Synthesized in the liver in the presence of vitamin K
- Converted to thrombin during coagulation

FACTOR III

- Tissue thromboplastin
- Released from damaged tissue
- Required to initiate the extrinsic cascade system

FACTOR IV

- Consists of calcium ions
- Required throughout the entire clotting sequence

FACTOR V

- Labile factor
- Synthesized in the liver
- Functions during the combined pathway phase of the coagulation system

FACTOR VII

- Serum prothrombin conversion accelerator or stable factor

- Synthesized in the liver in the presence of vitamin K
- Activated by factor III in the extrinsic system

FACTOR VIII

- Antihemophilic factor (antihemophilic globulin)
- Protein synthesized in the liver
- Required during the intrinsic phase of the coagulation system

FACTOR IX

- Plasma thromboplastin component
- Synthesized in the liver in the presence of vitamin K
- Required in the intrinsic phase of the coagulation system

FACTOR X

- Stuart factor (Stuart-Prower factor)
- Synthesized in the liver in the presence of vitamin K
- Required in the combined pathway of the coagulation system

FACTOR XI

- Plasma thromboplastin antecedent
- Synthesized in the liver
- Required in the intrinsic system

FACTOR XII

- Hageman factor
- Required in the intrinsic system

FACTOR XIII

- Fibrin stabilizing factor in the combined pathway phase of the coagulation system

Blood groups

- Determined by the presence or absence of genetically determined antigens or agglutinogens on the surface of RBCs
- A, B, and Rh are the most clinically significant blood antigens

ABO groups

- Type A blood has A antigen on its surface
- Type B blood has B antigen
- Type AB blood has both A and B antigens
- Type O blood has neither A nor B antigen
- Precise blood-typing and crossmatching (mixing and observing for agglutination of donor cells) are essential, especially for blood transfusions
- Donor's blood must be compatible with recipient's blood or the result can be fatal
  - Type O is universal donor
  - Type O can only receive type O

Rh typing

- Determines whether Rh factor is present or absent in the blood
  - Blood with the Rh antigen is Rh-positive
  - Blood without the Rh antigen is Rh-negative

Cross-training

Use your knowledge of hematologic terms to complete the crossword puzzle below.

Across

4. The process by which blood cells form and develop

5. Smallest of the white blood cells and the second most numerous

6. Substance secreted by basophils in response to certain inflammatory and immune stimuli

7. Another name for red blood cells

9. Another name for white blood cells

11. The oxygen-carrying substance that gives blood its red color

12. Cells that ingest microorganisms, cellular debris, and necrotic tissue

13. Substance on the surface of red blood cells that determines blood groups

Down

1. Decrease in the caliber of blood vessels

2. Play a crucial role in blood clotting

3. Immature red blood cell

4. Complex process by which platelets, plasma, and coagulation factors interact to control bleeding

8. Most numerous type of granulocyte

10. Largest of the white blood cells

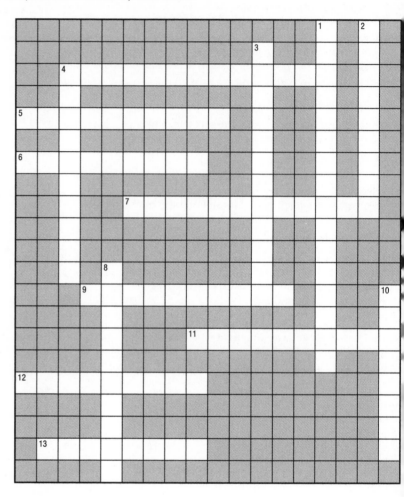

Parallel bars

Complete each of the analogies below by determining the relationship between the given word pair and then determining the missing term in the second word pair.

1. Platelet : _____ :: Red blood cell : Erythrocyte

2. _____ : Lymphocyte :: Granulocyte : Eosinophil

3. Red blood cell : Reticulocyte :: Neutrophil : _____

4. Factor I : _____ :: Factor XIII : Fibrin stabilizing factor

You'd never know it just to look at you, but you're a real powerhouse! I just found out that multipotential cells in your bone marrow give rise to five distinct cell types!

Strike out

Some of the following statements about red blood cells (RBCs) are incorrect. Cross out all of the incorrect statements.

1. RBCs transport oxygen and carbon dioxide to and from body tissues.

2. RBCs contain hemoglobin, the oxygen-carrying substance that gives blood its red color.

3. The type of hemoglobin in the RBC determines a person's blood type.

4. RBCs have an average life span of 12 months.

5. Reticulocytes mature into RBCs in about 1 day.

6. The spleen sequesters, or isolates, old, worn-out RBCs, removing them from circulation.

7. The rate of reticulocyte release is fixed from birth.

Power stretch

Unscramble the words on the left to reveal the names of the three types of granulocytes. Then draw lines from each of the boxes to the list on the right, linking each type to its particular characteristics.

A. Account for up to 7% of circulating WBCs

B. The most numerous of the granulocytes

C. Constitute fewer than 2% of circulating WBCs

HENSLIPTOUR

_ _ _ _ _ _ _ _ _ _ _

D. Form main component of pus

E. Accumulate in loose connective tissue

F. Leave the blood stream and migrate to and accumulate at infection sites

LIESHOOPINS

_ _ _ _ _ _ _ _ _ _ _

G. Possess little or no phagocytic ability

H. Engulf, ingest, and digest foreign materials

I. Migrate from the bloodstream as a response to an allergic reaction

ALBSHIPSO

_ _ _ _ _ _ _ _ _ _

J. Cytoplasmic granules secrete histamine

K. Replacements produced by bone marrow

L. Involved in the ingestion of antigen-antibody complexes

All granulocytes contain a single multilobular nucleus and granules in the cytoplasm.

But don't lump us all together! Each one of us exhibit different properties, and we're activated by different stimuli.

Coaching session
Comparing WBCs

Granulocytes
- *Neutrophils:* Engulf and digest foreign materials
- *Eosinophils:* Ingest antigens and antibodies
- *Basophils:* Secrete histamine in response to inflammatory and immune stimuli

Agranulocytes
- *Monocytes:* Ingest bacteria, cellular debris, and necrotic tissue
- *T lymphocytes:* Directly attack infected cells
- *B lymphocytes:* Produce antibodies against antigens

Hit or miss

Some of the following statements are true; the others are false. Mark each accordingly.

_____ 1. Plasma may contain antibodies that interact with the antigens on RBCs, causing the cells to agglutinate.

_____ 2. Testing for the presence of antibodies is the most important system for classifying blood.

_____ 3. Type A blood has A antibodies.

_____ 4. Of the eight types of Rh antigens, only C, D, and E are common.

_____ 5. Typically, blood does not contain the Rh antigen.

_____ 6. Patients with type O blood can receive all other blood types.

Match point

Match each of the terms on the left with its definition and characteristics on the right.

1. Intrinsic pathway _____

2. Extrinsic pathway _____

3. Factor I _____

4. Factor II _____

5. Factor III _____

6. Factor IV _____

7. Factor V _____

8. Factor VII _____

9. Factor VIII _____

10. Factor IX _____

11. Factor X _____

12. Factor XI _____

13. Factor XII _____

14. Factor XIII _____

A. Prothrombin; converted to thrombin during coagulation

B. Antihemophilic factor; required during the intrinsic phase of the coagulation system

C. Clotting pathway activated when plasma comes in contact with damaged vessel surfaces

D. Clotting pathway activated when tissue factor released by damaged tissue comes into contact with one of the clotting factors

E. Plasma thromboplastin antecedent; required in the intrinsic pathway

F. Serum prothrombin conversion accelerator or stable factor (proconvertin); activated by Factor III in the extrinsic system

G. Fibrin stabilizing factor; required to stabilize fibrin strands in the common pathway

H. Fibrinogen; converted to fibrin during coagulation cascade

I. Consists of calcium ions; required throughout the entire clotting sequence

J. Stuart factor (Stuart-Prower factor); required in the common pathway of the coagulation system

K. Hageman factor; required in the intrinsic pathway

L. Tissue factor (thromboplastin); released from tissue damage

M. Labile factor (proaccelerin); functions during the common pathway phase of the coagulation system

N. Plasma thromboplastin component; required in the intrinsic phase of the coagulation system

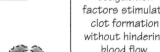

Coagulation factors stimulate clot formation without hindering blood flow.

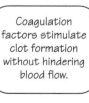

Interval training

Understanding blood types and compatibilities is important. Test your knowledge on the subject by completing the following chart. Note that a few clues are provided to get you started.

Blood group	Antibodies present in plasma	Compatible RBCs	Compatible plasma
Recipient			
O	_____ 1	_____ 8	O, A, B, AB
A	_____ 2	_____ 9	_____ 15
B	_____ 3	B, O	_____ 16
AB	Neither anti-A nor anti-B	_____ 10	_____ 17
Donor			
O	_____ 4	_____ 11	_____ 18
A	_____ 5	_____ 12	_____ 19
B	_____ 6	_____ 13	_____ 20
AB	_____ 7	_____ 14	_____ 21

9

Immune system

■■ ■ Warm-up

Immune system review

Immune system basics

■ Defends the body against invasion by harmful organisms or chemical toxins
■ Closely related to blood
■ Uses the bloodstream to transport defenses to the site of an invasion
■ Has three major divisions:
  – Central lymphoid organs and tissues
  – Peripheral lymphoid organs and tissues
  – Accessory lymphoid organs and tissues
■ Bone marrow and thymus play roles in development of B cells and T cells

Bone marrow

■ Site of immune system and blood cell development from stem cells (hematopoiesis)
■ Contains stem cells, which may develop into any of several different cell types and some serve as sources for lymphocytes and phagocytes

Lymphocytes

■ Further differentiated into B cells or T cells
■ B cells mature in the bone marrow
■ T cells travel to and mature in the thymus
■ Distributed throughout the lymphoid organs
■ Produce molecules called *antibodies* that attack pathogens or direct other cells, such as phagocytes, to attach them

Thymus

■ Helps form T lymphocytes for months after birth; then atrophies

Peripheral lymphoid organs and tissues

Lymph nodes

■ Small, oval-shaped structures located along a network of lymph channels
■ Most abundant in the head, neck, axillae, abdomen, pelvis, and groin
■ Help remove and destroy antigens that circulate in the blood and lymph

Lymph and lymphatic vessels

■ Clear fluid that bathes body tissues
■ Contains a liquid portion, white blood cells and antigens
■ Collected from body tissues, seeps into lymphatic vessels
■ Carried by afferent lymphatic vessels into the subcapsular sinus of the lymph node
■ When cleansed, leaves the node through efferent lymphatic vessels at the hilum
■ Axillary nodes filter drainage from the arms
■ Femoral nodes filter drainage from the legs

Spleen

■ Located in the left upper quadrant of the abdomen beneath the diaphragm
■ Dark red, oval structure that's approximately the size of a fist
■ White pulp—contains compact masses of lymphocytes surrounding branches of the splenic artery
■ Red pulp—consists of a network of blood-filled sinusoids—supported by a framework of reticular fibers and mononuclear phagocytes, along with some lymphocytes, plasma cells, and monocytes
■ Phagocytes
  – Engulf and break down worn-out red blood cells (RBCs)
  – Selectively retain and destroy damaged or abnormal RBCs
■ Filters and removes bacteria and other foreign substances that enter the bloodstream
■ Splenic phagocytes interact with lymphocytes to initiate an immune response
■ Stores blood and 20% to 30% of platelets

Accessory lymphoid organs and tissues

■ Tonsils, adenoids, appendix, and Peyer's patches
■ Remove foreign debris in much the same way lymph nodes do

Immune system function

■ *Immunity* refers to the body's capacity to resist invading organisms and toxins, thereby preventing tissue and organ damage
■ Designed to recognize, respond to, and eliminate antigens, including bacteria, fungi, viruses, and parasites
■ Preserves the body's internal environment by scavenging dead or damaged cells and patrolling for antigens

- Uses three basic strategies:
  - Protective surface phenomena
  - General host defenses
  - Specific immune responses

Protective surface phenomena

- Strategically placed physical, chemical, and mechanical barriers work to prevent the entry of potentially harmful organisms
- Intact and healing skin and mucous membranes provide the first line of defense against microbial invasion
- Skin desquamation (normal cell turnover) and low pH further impede bacterial colonization
- Seromucous surfaces are protected by antibacterial substances found in tears, saliva, and nasal secretions
- Nasal hairs and turbulent airflow through the nostrils filter foreign materials
- Nasal secretions contain an immunoglobulin that discourages microbe adherence
- A mucous layer lines the respiratory tract
- Bacteria are mechanically removed by saliva, swallowing, peristalsis, and defecation in the GI tract
- Low pH of gastric secretions renders the stomach virtually free from live bacteria
- Urinary system is sterile except for the distal end of the urethra and the urinary meatus
- Urine flow, low urine pH, immunoglobulin and, in men, the bactericidal effects of prostatic fluid work together to impede bacterial colonization
- A series of urinary sphincters also inhibits bacterial migration

General host defenses

- Inflammatory response eliminates dead tissue, microorganisms, toxins, and inert foreign matter
- Phagocytosis occurs after inflammation or during chronic infections

Specific immune responses

- Particular microorganisms or molecules activate specific immune responses
- Specific responses (*humoral immunity* or *cell-mediated immunity*) are produced by lymphocytes

HUMORAL IMMUNITY

- Invading antigen causes B cells to divide and differentiate into plasma cells
- Each plasma cell produces and secretes large amounts of antigen-specific immunoglobulins into the bloodstream
- Each of the five types of immunoglobulin serves a particular function:
  - IgA, IgG, and IgM guard against viral and bacterial invasion
  - IgD acts as an antigen receptor of B cells
  - IgE causes an allergic response

- Depending on the antigen, immunoglobulins can work in one of several ways:
  - Disable certain bacteria by linking with toxins that the bacteria produce; these immunoglobulins are called *antitoxins*
  - Opsonize (coat) bacteria, making them targets for scavenging by phagocytosis
  - Link to antigens, causing the immune system to produce and circulate enzymes called *complement*
- Primary antibody response occurs 4 to 10 days after first-time antigen exposure
  - Immunoglobulin levels increase, then quickly dissipate, and IgM antibodies form
- Subsequent exposure to the same antigen initiates a secondary antibody response
  - Memory B cells manufacture antibodies achieving peak levels in 1 to 2 days but persist for months, then fall slowly
  - Faster, more intense, and more persistent than the primary response
- Response intensifies with each subsequent exposure to the same antigen
- After the antibody reacts to the antigen, an antigen-antibody complex forms
- Then the antibody activates the complement system, causing an enzymatic cascade that destroys the antigen

COMPLEMENT SYSTEM

- Bridges humoral and cell-mediated immunity
- Attracts phagocytic neutrophils and macrophages to the antigen site
- Consists of about 25 enzymes that "complement" the work of antibodies by aiding phagocytosis or destroying bacteria cells
- Complement proteins travel in the bloodstream in an inactive form
- When the first complement substance is triggered, the complement cascade is set in motion

COMPLEMENT CASCADE

- Leads to the creation of the membrane attack complex
- Creates a channel through which fluids and molecules flow in and out
- Target cell swells and eventually bursts
- By-products of the complement cascade also produce:
  - Inflammatory response
  - Stimulation and attraction of neutrophils
  - Coating of target cells by C3b (an inactivated fragment of the complement protein C3), making them attractive to phagocytes

CELL-MEDIATED IMMUNITY

- Protects the body against bacterial, viral, and fungal infections
- Provides resistance against transplanted cells and tumor cells

- A macrophage processes the antigen, which is then presented to T cells
- Some T cells become sensitized and destroy the antigen; other T-cells release lymphokines, which activate macrophages that destroy the antigen
- Sensitized T cells then travel through the blood and lymphatic system and provide ongoing surveillance in their quest for specific antigens

Immune system failure

Hypersensitivity disorders

- Caused by an exaggerated or inappropriate immune response
- Such disorders are classified as type I through type IV

TYPE I DISORDERS

- Anaphylactic (immediate, atopic, IgE-mediated reaginic) reactions
- Examples: Systemic anaphylaxis, hay fever (seasonal allergic rhinitis), reactions to insect stings, some food and drug reactions, some cases of urticaria, infantile eczema

TYPE II DISORDERS

- Cytotoxic (cytolytic, complement-dependent cytotoxicity) reactions
- Examples: Goodpasture's syndrome, autoimmune hemolytic anemia, transfusion reactions, hemolytic disease of the newborn, myasthenia gravis, some drug reactions

TYPE III DISORDERS

- Immune complex disease reactions
- Examples: Reactions associated with such infections as hepatitis B and bacterial endocarditis; cancers, in which a serum sickness-like syndrome may occur; autoimmune disorders such as systemic lupus erythematosus
- May also follow drug or serum therapy

TYPE IV DISORDERS

- Delayed (cell-mediated) hypersensitivity reactions
- Examples: Tuberculin reactions, contact hypersensitivity, sarcoidosis

Autoimmune disorders

- Marked by an abnormal response to one's own tissue
- Examples: Rheumatoid arthritis, psoriatic arthritis, ankylosing spondylitis, Sjögren's syndrome, lupus erythematosus

Immunodeficiency

- Caused by an absent or depressed immune response in various forms
- Examples: X-linked infantile hypogammaglobulinemia, common variable immunodeficiency, DiGeorge's syndrome, acquired immunodeficiency syndrome, chronic granulomatous disease, ataxia-telangiectasia, severe combined immunodeficiency disease, complement deficiencies

■ Finish line

This illustration shows central, peripheral, and accessory lymphoid organs and tissue. See if you can name them all.

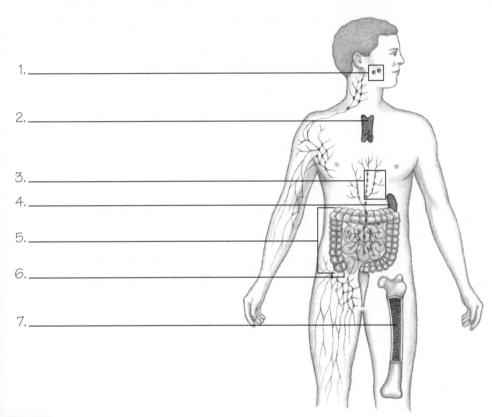

1. _____

2. _____

3. _____

4. _____

5. _____

6. _____

7. _____

■ Train your brain

Sound out each group of pictures and symbols to uncover a message about the central immune system.

Cross-training

Test your knowledge of terminology related to the immune system by completing the crossword puzzle below.

Across

4. Type of disorder resulting from an exaggerated or inappropriate immune response

7. Receptors that attack pathogens or direct other cells, such as phagocytes, to attack for them

9. Type of immunoglobulin that can disable certain bacteria by linking with toxins that the bacteria produce

10. Process by which cells of the immune system and the blood develop from stem cells

11. The body's capacity to resist invading organisms and toxins

12. Largest lymphatic organ

13. Cells that ingest organisms

Down

1. Clear fluid that bathes the body tissues

2. Cells capable of taking many forms

3. Two-lobed mass of lymphoid tissue located in the mediastinum

5. Type of disorder caused by an absent or depressed immune response

6. The most numerous polymorphonuclear leukocytes

7. Substance capable of triggering an immune response

8. Type of disorder marked by an abnormal response to one's own tissue

■ Strike out

Some of the following statements about the immune system are incorrect. Cross out all of the incorrect statements.

1. The bone marrow contains stem cells that can develop into any of several different cell types.

2. Some of these cells develop into lymphocytes; other cells develop into phagocytes.

3. Those that become phagocytes are further differentiated to become either B cells or T cells.

4. B cells remain in the bone marrow to mature; T cells travel to the thymus and mature there.

5. Once mature, B cells and T cells remain in either the bone marrow or the thymus.

6. B cells have immunoglobulins, or antibodies, that can attack other pathogens.

7. The thymus remains active throughout adulthood.

8. In the thymus, T cells undergo T-cell education, in which the cells are trained to differentiate between cells from the same body and all other cells.

Memory jogger

To recall where lymphocytes mature, think "B in B and T in T":

1. B cells mature in the
   B_____.

2. T cells mature in the
   T_____.

While we may look like just a bunch of Ts, we're not all alike. Each type of T cell has a specific function.

Match point

Match each of the terms on the left with its definition on the right.

1. Lymph nodes _____
2. Superficial cortex _____
3. Deep cortex _____
4. Medulla _____
5. Lymph _____
6. Lymphatic vessels _____
7. Afferent lymphatic vessels _____
8. Efferent lymphatic vessels _____
9. White pulp _____
10. Red pulp _____

A. Compartment in lymph nodes consisting mostly of T cells

B. Portion of the spleen's interior containing compact masses of lymphocytes surrounding branches of the splenic artery

C. Oval-shaped structures located along a network of lymph channels

D. Thin-walled drainage channels connecting lymphatic tissues

E. Portion of the spleen's interior containing a network of blood-filled sinusoids

F. Compartment in lymph nodes containing follicles made up predominantly of B cells

G. Compartment in lymph nodes containing numerous plasma cells that actively secrete immunoglobulins

H. Carry cleansed lymph away from the lymph node

I. Clear fluid containing WBCs and antigens that bathes body tissues

J. Carry lymph into the subcapsular sinus of the lymph node

Finish line

Identify the structures of the lymph node shown in the illustration below.

1. _____
2. _____
3. _____
4. _____
5. _____
6. _____
7. _____
8. _____
9. _____
10. _____

Lymphatic capillaries are located throughout most of the body. Wider than blood capillaries, they permit interstitial fluid to flow into them but not out.

A-maze-ing race

Mr. Spleen has a lot of tasks to complete to ensure that the body functions properly. Help him find his way through this maze by correctly answering "true" or "false" to each of the following statements about the spleen.

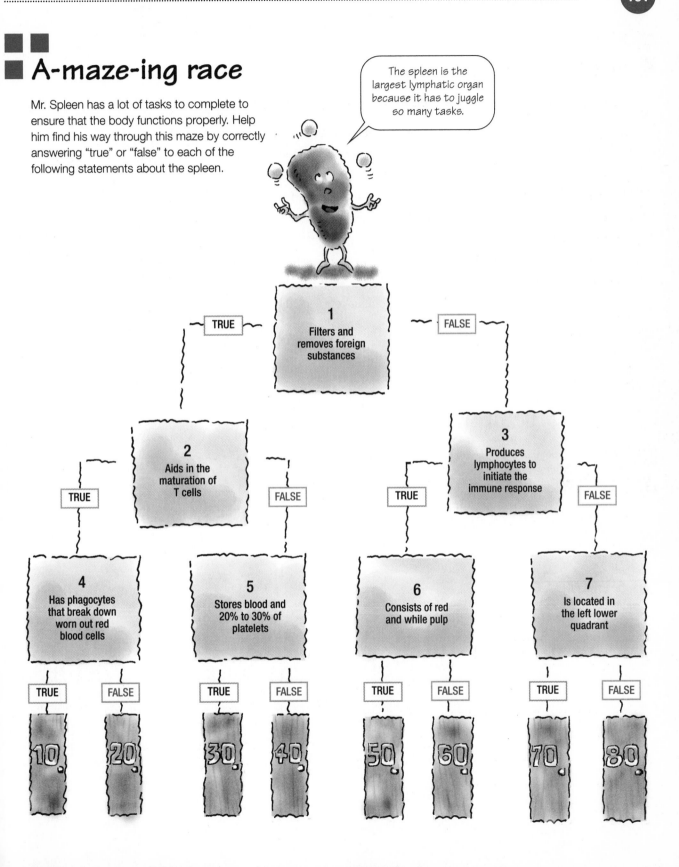

The spleen is the largest lymphatic organ because it has to juggle so many tasks.

TRUE

FALSE

1
Filters and removes foreign substances

2
Aids in the maturation of T cells

3
Produces lymphocytes to initiate the immune response

TRUE

FALSE

TRUE

FALSE

4
Has phagocytes that break down worn out red blood cells

5
Stores blood and 20% to 30% of platelets

6
Consists of red and while pulp

7
Is located in the left lower quadrant

TRUE **FALSE** **TRUE** **FALSE** **TRUE** **FALSE** **TRUE** **FALSE**

10 20 30 40 50 60 70 80

■ Hit or miss

Some of the following statements are true; the others are false. Mark each accordingly.

_____ 1. The respiratory system provides the first line of defense against microbial invasion.

_____ 2. Skin desquamation (normal cell turnover) places skin at risk for bacterial invasion.

_____ 3. The enzyme lysozyme, which is found in tears, saliva, and nasal secretions, is antibacterial.

_____ 4. The easiest part of the body for microorganisms to enter is the respiratory system.

_____ 5. Nasal hairs and turbulent airflow filter out foreign materials.

_____ 6. Moist nasal secretions create an ideal environment for bacterial growth.

_____ 7. In the GI tract, bacteria are mechanically removed by saliva, swallowing, peristalsis, and defecation.

_____ 8. The stomach and intestines have few defenses against invasion by bacteria.

_____ 9. Urine flow, low urine pH, and immunoglobulin help keep the urinary system sterile.

■ Circuit training

Trace the sequence of events in the inflammatory response by linking the boxes below with arrows, beginning with the invasion of tissue by microorganisms and ending with repaired tissue.

| Vasodilation occurs along with increased capillary permeability. |

| Microorganisms invade damaged tissue. |

| Basophils release heparin, and histamine and kinin production occurs. |

| Blood flow increases to the affected tissue and fluid collects within it. |

| Tissue is repaired. |

| Neutrophils flock to the invasion site to engulf and destroy microorganisms from dying cells. |

■ Power stretch

Unscramble the words on the left to reveal the names of the major polymorphonuclear leukocytes involved in the inflammatory response. Then draw lines from the boxes on the left to the list on the right, linking each leukocyte to its particular characteristics. (*Note:* Some characteristics may link to more than one leukocyte.)

A. Increase dramatically in number in response to infection and inflammation

B. Circulate in peripheral blood

C. Are attracted to areas of inflammation

D. Multiply in allergic and parasitic disorders

E. Accumulate in connective tissue (particularly in the lungs, intestines, and skin)

F. Have surface receptors for immunoglobulin (Ig) E

G. Most numerous polymorphonuclear leukocytes

H. Participate in host defense against parasites

I. When their receptors are cross-linked by an IgE antigen complex, release mediators characteristic of the allergic response

J. Engulf, digest, and dispose of invading organisms through phagocytosis

K. Main constituent of pus

SOUTHERNLIP

— — — — — — — — — — —

PHONEOILSIS

— — — — — — — — — — —

ABISHOPSL

— — — — — — — — —

STAM ELLCS

— — — —   — — — — —

Pep talk

"We are what we repeatedly do. Excellence, therefore, is not an act but a habit.

—Aristotle

■ Match point

Match each of the terms on the left with its definition on the right.

1. Humoral immunity _____

2. Antitoxins _____

3. Complement _____

4. Primary antibody response _____

5. Secondary antibody response _____

6. Antigen-antibody complex _____

7. Complement system _____

8. Cell-mediated immunity _____

9. Lymphokines _____

10. Cytokines _____

11. Membrane attack complex _____

A. Response occurring 4 to 10 days after first-time antigen exposure

B. Specific immune response in which T cells move directly to attack invading antigens

C. Specific immune response in which antigens cause B cells to differentiate into plasma cells and produce circulating immunoglobulins

D. Complex inserted into the membrane of the target cell that creates a channel for fluids and molecules

E. Immunoglobulins that link with toxins produced by bacteria

F. Forms after an antibody reacts to an antigen

G. Enzymes produced and circulated by the immune system after immunoglobulins link to antigens

H. Proteins involved in the communication between macrophages and lymphocytes

I. Response achieving peak levels in 1 to 2 days

J. Bridges humoral and cell-mediated immunity

K. Substance secreted by T cells, which activates macrophages to destroy an antigen

All foreign substances elicit the same general host defenses. In addition, particular microorganisms activate specific immune responses.

You make the call

Using the illustrations below as a guide, explain the five steps in phagocytosis.

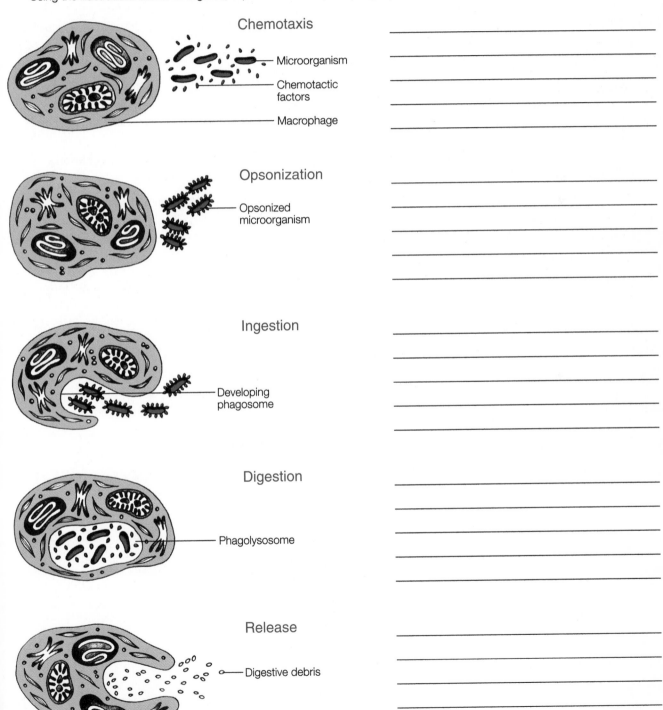

Chemotaxis

— Microorganism

— Chemotactic factors

— Macrophage

_____
_____
_____
_____
_____

Opsonization

— Opsonized microorganism

_____
_____
_____
_____
_____
_____

Ingestion

— Developing phagosome

_____
_____
_____
_____
_____
_____

Digestion

— Phagolysosome

_____
_____
_____
_____
_____
_____

Release

— Digestive debris

_____
_____
_____
_____
_____
_____

Strike out

Some of the following statements are incorrect. Cross out all of the incorrect statements.

1. T cells are involved in humoral immunity.

2. IgA, IgG, and IgM guard against viral and bacterial invasion.

3. IgE acts as an antigen receptor of B cells.

4. All immunoglobulins work to disable bacteria by producing antitoxins.

5. During the primary antibody response, immunoglobulin levels increase, then quickly dissipate, and IgM antibodies form.

6. During a secondary antibody response, memory B cells again manufacture IgM antibodies, this time achieving peak levels in 1 to 2 days.

7. The complement system consists of about 25 enzymes that aid phagocytosis or destroy bacteria cells.

8. The complement cascade occurs when complement substances are triggered in a sequence of controlled steps, leading to the creation of the membrane attack complex.

9. One of the by-products of the complement cascade is to produce more T cells.

10. In cell-mediated immunity, some T cells become sensitized and destroy the antigen; others release lymphokines, which activate macrophages that destroy the antigen.

Match point

Match each disorder listed on the left with an example of that disorder, listed on the right.

1. Anaphylactic reaction (type I hypersensitivity disorder) _____

2. Cytotoxic reaction (type II hypersensitivity disorder) _____

3. Immune complex disease reaction (type III hypersensitivity disorder) _____

4. Delayed hypersensitivity reaction (type IV hypersensitivity disorder) _____

5. Autoimmune disorder _____

6. Immunodeficiency disorder _____

A. Contact hypersensitivity

B. Hay fever

C. X-linked infantile hypogammaglobulinemia

D. Autoimmune hemolytic anemia

E. Reactions associated with hepatitis B

F. Rheumatoid arthritis

10

Respiratory system

Respiratory system review

Respiratory system basics

■ Maintains the exchange of oxygen and carbon dioxide in the lungs and tissues
■ Helps regulate the body's acid-base balance

Upper respiratory tract

■ Consists primarily of the nose, mouth, nasopharynx, oropharynx, laryngopharynx, and larynx
■ Warms and humidifies inspired air
■ Responsible for taste, smell, and the chewing and swallowing of food

Nostrils and nasal passages

■ Air enters the body through the nostrils (nares)
■ In the nares, small hairs (vibrissae) filter out dust and large foreign particles
■ Air then passes into the two nasal passages, which are separated by the septum
■ Cartilage forms the anterior walls of the nasal passages
■ Bony structures (conchae or turbinates) form the posterior walls
■ Conchae warm and humidify air before it passes into the nasopharynx
■ Mucus layer also traps finer foreign particles, which the cilia carry to the pharynx to be swallowed

Sinuses and pharynx

■ Four paranasal sinuses are located in the frontal, sphenoid, and maxillary bones
■ Sinuses provide speech resonance
■ Air passes from the nasal cavity into the muscular nasopharynx through the choanae—a pair of posterior openings in the nasal cavity that remain constantly open

Oropharynx and laryngopharynx

■ Oropharynx (in posterior wall of the mouth) connects the nasopharynx and the laryngopharynx
■ Laryngopharynx extends to the esophagus and larynx

Larynx

■ Contains the vocal cords
■ Connects the pharynx with the trachea
■ Muscles and cartilage form the walls of the larynx, including the large, shield-shaped thyroid cartilage situated just under the jaw line

Lower respiratory tract

■ Consists of the trachea, bronchi, and lungs
■ Subdivided into the conducting airways and the acini
■ Acinus serves as the area of gas exchange
■ Mucous membrane contains hairlike cilia and lines the lower tract
■ Cilia constantly clean the tract and carry foreign matter upward for swallowing or expectoration

Conducting airways

■ Contain the trachea and bronchi
■ Help facilitate gas exchange

TRACHEA

■ Extends from the cricoid cartilage at the top to the carina (also called the *tracheal bifurcation*)
– Carina: A ridge-shaped structure at the level of the sixth or seventh thoracic vertebra
■ Protected by C-shaped cartilage rings that reinforces it and prevents it from collapsing

BRONCHI

■ Primary bronchi begin at the carina
– Right mainstem bronchus is shorter, wider, and more vertical than the left; supplies air to the right lung
– Left mainstem bronchus delivers air to the left lung
■ Mainstem bronchi divide into the five lobar bronchi (secondary bronchi)
■ Along with blood vessels, nerves, and lymphatics, the secondary bronchi enter the pleural cavities and the lungs at the hilum (a slit located behind the heart on the lung's medial surface where the lungs are anchored)
■ Each lobar bronchus enters a lobe in each lung
■ Within its lobe, each of the lobar bronchi branches into segmental bronchi (tertiary bronchi)
■ Segments continue to branch into smaller and smaller bronchi, finally branching into bronchioles
■ Larger bronchi consist of cartilage, smooth muscle, and epithelium
■ As bronchi become smaller, they first lose cartilage, then smooth muscle until, finally, the smallest bronchioles consist of just a single layer of epithelial cells

Acinus

- Each bronchiole includes terminal bronchioles and the acinus
- Chief respiratory unit for gas exchange

RESPIRATORY BRONCHIOLES

- Branch off of terminal bronchioles
- Feed directly into alveoli at sites along their walls

ALVEOLAR DUCTS

- Branch off of respiratory bronchioles
- Terminate in clusters of capillary-swathed alveoli called *alveolar sacs*

ALVEOLI

- Site of gas exchange
- Walls contain two basic epithelial cell types
  - Type I: most abundant; thin, flat, squamous cells across which gas exchange occurs
  - Type II: secrete surfactant, a substance that coats the alveolus and promotes gas exchange by lowering surface tension

Lungs and accessory structures

- Cone-shaped organs that hang suspended in the right and left pleural cavities, straddling the heart, and anchored by root and pulmonary ligaments
- Right lung (shorter, broader, and larger than the left) has three lobes and handles 55% of gas exchange
- Left lung has two lobes
- Each lung's concave base rests on the diaphragm and apex extends about ½″ (1.3 cm) above the first rib

Pleura and pleural cavities

PLEURA

- Membrane that totally encloses the lung
- Composed of a visceral layer and a parietal layer
  - Visceral pleura hugs the entire lung surface, including the areas between the lobes
  - Parietal pleura lines the inner surface of the chest wall and upper surface of the diaphragm

PLEURAL CAVITY

- Tiny area between the visceral and parietal pleural layers
- Contains a thin film of serous fluid
- Fluid has two functions:
  - Lubricates the pleural surfaces so that they slide smoothly against each other as the lungs expand and contract
  - Creates a bond between the layers that causes the lungs to move with the chest wall during breathing

Thoracic cavity

- Area surrounded by the diaphragm (below), the scalene muscles and fasciae of the neck (above), and the ribs, intercostal muscles, vertebrae, sternum, and ligaments (around the circumference)

Mediastinum

- Space between the lungs
- Contains:
  - Heart and pericardium
  - Thoracic aorta
  - Pulmonary artery and veins
  - Venae cavae and azygos veins
  - Thymus, lymph nodes, and vessels
  - Trachea, esophagus, and thoracic duct
  - Vagus, cardiac, and phrenic nerves

Thoracic cage

- Composed of bone and cartilage
- Supports and protects the lungs
- Allows the lungs to expand and contract

POSTERIOR THORACIC CAGE

- Contains vertebral column and 12 pairs of ribs form the posterior portion of the thoracic cage
- The ribs form the major portion of the thoracic cage
- They extend from the thoracic vertebrae toward the anterior thorax

ANTERIOR THORACIC CAGE

- Consists of the manubrium, sternum, xiphoid process, and ribs
- Protects the mediastinal organs that lie between the right and left pleural cavities
- Ribs 1 through 7 attach directly to the sternum
- Ribs 8 through 10 attach to the cartilage of the preceding rib
- The other 2 pairs of ribs are "free-floating"—they don't attach to any part of the anterior thoracic cage
- Rib 11 ends anterolaterally
- Rib 12 ends laterally
- Lower parts of the rib cage (costal margins) near the xiphoid process form the borders of the costal angle—an angle of about 90 degrees in a normal person
- Above the anterior thorax is a depression called the *suprasternal notch*

Respiration

- Composed of two processes
  - Inspiration—an active process
  - Expiration—a relatively passive process
  - Both actions rely on respiratory muscle function and the effects of pressure differences in the lungs

- During normal respiration, the external intercostal muscles (located between and along the lower borders of the ribs) aid the diaphragm (major muscle of respiration)
- The diaphragm descends to lengthen the chest cavity, while the external intercostal muscles contract to expand the anteroposterior diameter
- This coordinated action causes inspiration
- Rising of the diaphragm and relaxation of the intercostal muscles causes expiration

Forced inspiration and active expiration

- During exercise, when the body needs increased oxygenation, or in certain disease states that require forced inspiration and active expiration, the accessory muscles of respiration also participate

FORCED INSPIRATION

- Pectoral muscles (upper chest) raise the chest to increase the anteroposterior diameter
- Sternocleidomastoid muscles (side of neck) raise the sternum
- Scalene muscles (in the neck) elevate, fix, and expand the upper chest
- Posterior trapezius muscles (upper back) raise the thoracic cage

ACTIVE EXPIRATION

- Internal intercostal muscles contract to shorten the chest's transverse diameter
- Abdominal rectus muscles pull down the lower chest, thus depressing the lower ribs
- Oxygen-depleted blood enters the lungs from the pulmonary artery, then flows through the main pulmonary arteries into the smaller vessels of the pleural cavities and the main bronchi through the arterioles and, eventually, to the capillary networks in the alveoli
- In the alveoli, gas exchange (oxygen and carbon dioxide diffusion) takes place

External and internal respiration

- Effective respiration consists of gas exchange in the lungs, called *external respiration*, and gas exchange in the tissues, called *internal respiration*
- External respiration occurs through three processes:
  - Ventilation—gas distribution into and out of the pulmonary airways
  - Pulmonary perfusion—blood flow from the right side of the heart, through the pulmonary circulation, and into the left side of the heart
  - Diffusion—gas movement through a semipermeable membrane from an area of greater to lesser concentration
- Internal respiration occurs only through diffusion

VENTILATION

- Distribution of gases (oxygen and carbon dioxide) into and out of the pulmonary airways
- Problems within the nervous, musculoskeletal, and pulmonary systems compromise breathing effectiveness
- Involuntary breathing results from stimulation of the respiratory center in the medulla and the pons of the brain
- Medulla controls the rate and depth of respiration; the pons moderates the rhythm of the switch from inspiration to expiration
- Adult thorax is flexible—its shape can be changed by contracting the chest muscles
- Medulla controls ventilation primarily by stimulating contraction of the diaphragm and external intercostal muscles, producing intrapulmonary pressure changes that cause inspiration
- Airflow distribution can be affected by many factors:
  - Airflow pattern
  - Volume and location of the functional reserve capacity (air retained in the alveoli that prevents their collapse)
  - Degree of intrapulmonary resistance
  - Presence of lung disease
- If airflow is disrupted for any reason, airflow distribution follows the path of least resistance
- Other musculoskeletal and intrapulmonary factors can affect airflow and, in turn, may affect breathing
- Other airflow alterations can also increase oxygen and energy demand and cause respiratory muscle fatigue
  - Interference with expansion of the lungs or thorax (changes in compliance)
  - Interference with airflow in the tracheobronchial tree (changes in resistance)

PULMONARY PERFUSION

- Blood flow from the right side of the heart, through the pulmonary circulation, and into the left side of the heart
- Aids external respiration
- Normal pulmonary blood flow allows alveolar gas exchange
- Many factors may interfere with gas transport to the alveoli
  - Cardiac output less than the average of 5 L/minute decreases gas exchange by reducing blood flow
  - Elevations in pulmonary and systemic resistance reduce blood flow
  - Abnormal or insufficient hemoglobin picks up less oxygen for exchange
- Gravity can promote oxygen and carbon dioxide transport by causing more unoxygenated blood to travel to the lower and middle lung lobes than to the upper lobes
- Gas exchange is most efficient in areas where perfusion and ventilation are similar (ventilation-perfusion match)

OPTIMAL CONDITIONS FOR DIFFUSION

- Oxygen and carbon dioxide molecules move between the alveoli and capillaries
- Direction of movement is always from an area of greater concentration to one of lesser concentration
- Oxygen moves across the alveolar and capillary membranes, dissolves in the plasma, and then passes through the red blood cell (RBC) membrane
- Carbon dioxide moves in the opposite direction
- The epithelial membranes lining the alveoli and capillaries must be intact
- Both the alveolar epithelium and the capillary endothelium are composed of a single layer of cells
- Between these layers are tiny interstitial spaces filled with elastin and collagen
- Normally, oxygen and carbon dioxide move easily through all of these layers
- Oxygen moves from the alveoli into the bloodstream, where it's taken up by hemoglobin in the RBCs
- When oxygen arrives in the bloodstream, it displaces carbon dioxide (the by-product of metabolism), which diffuses from RBCs into the blood and then to the alveoli
- Most transported oxygen binds with hemoglobin to form oxyhemoglobin, but a small portion dissolves in the plasma
- The portion of oxygen dissolved in plasma can be measured as the partial pressure of oxygen in arterial blood, or PaO_2
- After oxygen binds to hemoglobin, RBCs travel to the tissues
- Through cellular diffusion, internal respiration occurs when RBCs release oxygen and absorb carbon dioxide
- The RBCs then transport the carbon dioxide back to the lungs for removal during expiration

Acid-base balance

- Oxygen taken up in the lungs is transported to the tissues by the circulatory system
- Circulatory system exchanges it for carbon dioxide produced by metabolism in body cells
- Because carbon dioxide is more soluble than oxygen, it dissolves in the blood
- In the blood, most of the oxygen forms bicarbonate (base) and smaller amounts form carbonic acid (acid)

Respiratory responses

- The lungs control bicarbonate levels by converting bicarbonate to carbon dioxide and water for excretion
- In response to signals from the medulla, the lungs can change the rate and depth of breathing
- This change allows for adjustments in the amount of carbon dioxide lost to help maintain acid-base balance

METABOLIC ALKALOSIS

- Excess bicarbonate retention
- Rate and depth of ventilation decrease so that carbon dioxide can be retained, thus increasing carbonic acid levels

METABOLIC ACIDOSIS

- Excess acid retention or excess bicarbonate loss
- The lungs increase the rate and depth of ventilation to eliminate excess carbon dioxide, thus reducing carbonic acid levels
- An acid-base imbalance results when the lungs don't function properly

Batter's box

Fill in the blanks with the appropriate words. *Hint:* Some answers are used more than once.

Respiratory system essentials

The respiratory system consists of the _____ 1 respiratory tract, the

_____ 2 respiratory tract, including the _____ 3 and accessory

structures, and the _____ 4 cavity. The upper respiratory tract consists

primarily of the _____ 5 (nostrils and nasal passages), _____ 6 ,

nasopharynx, oropharynx, laryngopharynx, and _____ 7 . The lower

respiratory tract consists of the _____ 8 , _____ 9 , and

_____ 10 . The respiratory system maintains the exchange of

_____ 11 and _____ 12 in the lungs and tissues. It also helps

regulate the body's _____ 13 balance.

Options
acid-base

bronchi

carbon dioxide

larynx

lower

lungs

mouth

nose

oxygen

thoracic

trachea

upper

Sure, most people know that the upper respiratory tract filters, warms, and humidifies air. But did you also know that it's responsible for detecting taste and smell and chewing and swallowing food?

Cross-training

Test your knowledge of terms related to the respiratory system by completing this crossword puzzle.

Across

1. Structure that warms and humidifies air before it passes into the nasopharynx
3. Contains the vocal cords and connects the pharynx with the trachea
5. Reduced rate and depth of ventilation
7. Small hairs inside the nasal passages
9. Separates the two nasal passages
10. Ridge-shaped structure at the level of the sixth or seventh thoracic vertebrae
12. The posterior wall of the mouth; connects the nasopharynx and the laryngopharynx

Down

2. Increased rate and depth of ventilation
4. Chief respiratory unit for gas exchange
5. Slit on the lung's medial surface where secondary bronchi enter the pleural cavities
6. Membrane that totally encloses the lung
8. Substance that coats the alveolus and promotes gas exchange by lowering surface tension
11. Nasal passages through which air enters the respiratory system

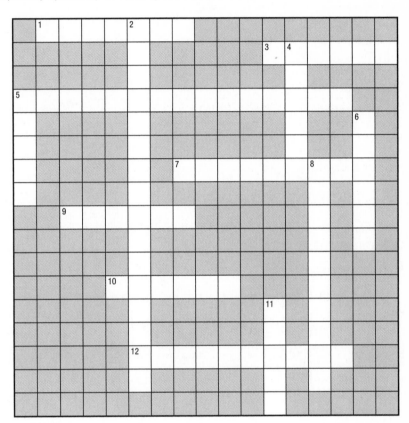

■■
■ Finish line

Identify the structures of the upper respiratory tract and the lower respiratory tract indicated on this illustration.

Upper respiratory tract

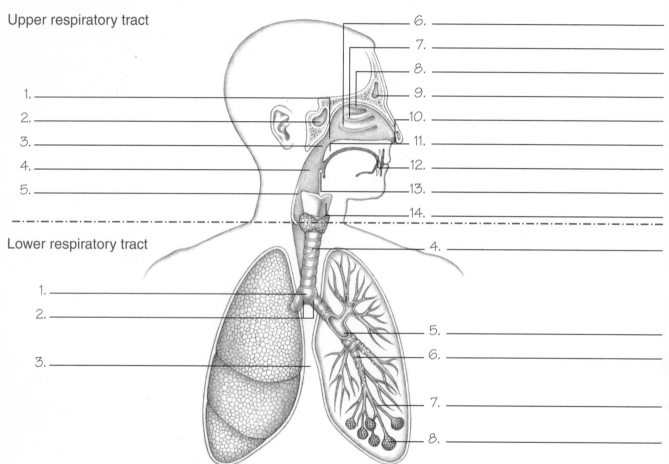

6. _____

7. _____

8. _____

1. _____
9. _____

2. _____
10. _____

3. _____
11. _____

4. _____
12. _____

5. _____
13. _____

14. _____

Lower respiratory tract

4. _____

1. _____

2. _____

5. _____

6. _____

3. _____

7. _____

8. _____

Train your brain

Sound out each group of pictures and symbols to reveal information about one aspect of the respiratory system.

Jumble gym

Use the clues to help you unscramble the following words related to the respiratory system. Then use the circled letters to answer the question posed.

Question: What's the name of the pair of posterior openings in the nasal cavity that remain constantly open?

1. Forms the anterior walls of the nasal passages

   R A C E L A G I T _ _ _ _ _ _ ◯ _ _

2. Small hairlike projections that help clean the respiratory tract

   C L A I I _ _ _ _ ◯

3. Bony structures forming the posterior walls of the nasal cavity

   C A N E C H O _ _ _ _ ◯ _ _

4. Portion of the lower respiratory tract that distributes air to the respiratory system

   C U D C O N T I N G S A R I Y A W ◯ _ _ _ _ _ _ _ ◯ _ _ _ _ _ _ _ _

5. Landmark at the top of the trachea

   C R I D O I C E C L A I R T A G _ _ _ _ ◯ _ _ _ _ _ _ _ _ _ _ ◯

Answer: _ _ _ _ _ _ _

■■ ■ Hit or miss

Some of the following statements about respiratory structures are true; the others are false. Mark each accordingly.

_____ 1. The right mainstem bronchus is shorter, wider, and more vertical than the left.

_____ 2. The mainstem bronchi divide into five lobar bronchi, also called secondary bronchi.

_____ 3. The secondary bronchi enter the lungs at the carina.

_____ 4. Each lobar bronchus enters a lobe in each lung.

_____ 5. The right lung has two lobes.

_____ 6. The lobar bronchi continue without branching until they reach the alveolus.

_____ 7. The larger bronchi consist entirely of cartilage.

_____ 8. The smallest bronchioles consist of just a single layer of epithelial cells.

_____ 9. Within the acinus, terminal bronchioles branch into smaller respiratory bronchioles, which feed directly into alveoli.

_____ 10. Alveolar walls contain two basic epithelial cell types.

_____ 11. Type I cells, which are the most abundant, secrete surfactant.

_____ 12. The respiratory bronchioles terminate in alveolar sacs.

_____ 13. Gas exchange occurs through the alveoli.

■■ ■ Finish line

Identify the structures of the pulmonary airway shown here.

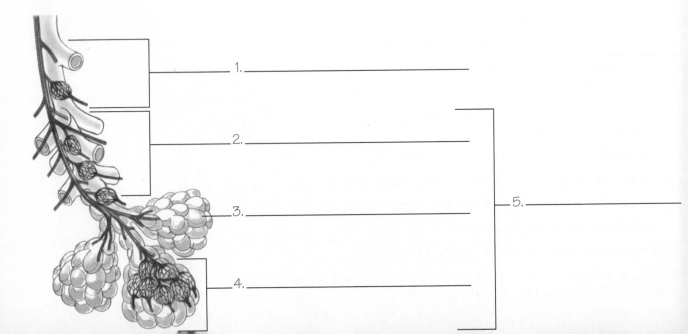

1. _____

2. _____

3. _____

4. _____

5. _____

■ Match point

Match each of the terms below with its definition.

1. Pleura _____

2. Visceral pleura _____

3. Parietal pleura _____

4. Pleural cavity _____

5. Serous fluid _____

6. Thoracic cavity _____

7. Mediastinum _____

8. Thoracic cage _____

9. Suprasternal notch _____

A. Area above the anterior thorax where trachea and aortic pulsation can be palpated

B. Area surrounded by the diaphragm (below), the scalene muscles and fasciae of the neck (above), and the ribs (around the circumference)

C. Membrane that totally encloses the lung

D. Covering that hugs the entire lung surface, including the areas between the lobes

E. Area composed of bone and cartilage that supports and protects the lungs

F. Fluid that fills the pleural cavity

G. Covering that lines the inner surface of the chest wall and upper surface of the diaphragm

H. Tiny area between the visceral and parietal pleural layers

I. Space between the lungs

■ Strike out

Some of the following statements about the respiratory system structures are incorrect. Cross out all of the incorrect statements.

1. The left lung is larger than the right and handles the majority of gas exchange.

2. One function of serous fluid is to lubricate the pleural surfaces, allowing them to slide smoothly against each other as the lungs expand and contract.

3. Another function of serous fluid is to prevent inflammation on the lung surface.

4. Each lung's base rests on the diaphragm.

5. Cartilage forms the major portion of the thoracic cage.

6. The anterior thoracic cage consists of the manubrium, sternum, xiphoid process, and ribs.

7. The posterior thoracic cage is composed of the vertebral column and 12 pairs of ribs.

8. All of the ribs attach directly to the sternum.

9. The costal margin is the lower parts of the rib cage.

10. The costal angle is the angle of about 90 degrees that forms at the point where the bottom rib meets the 12th thoracic vertebra.

Finish line

Identify the structures of the thoracic cage indicated in these illustrations.

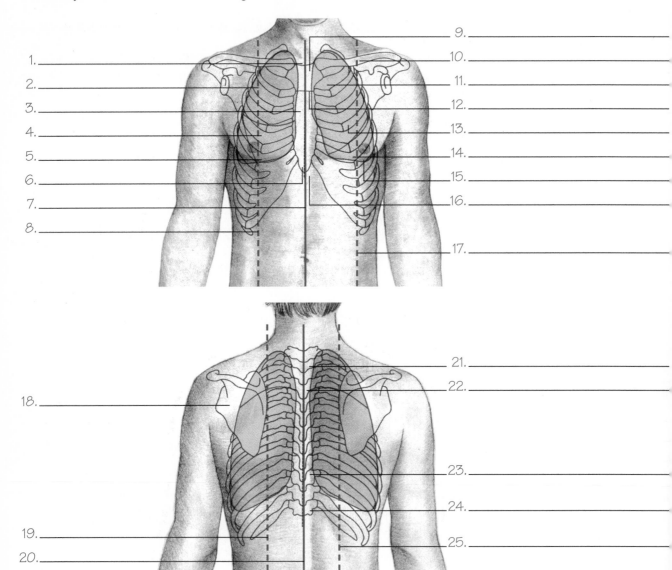

1. _____
2. _____
3. _____
4. _____
5. _____
6. _____
7. _____
8. _____

9. _____
10. _____
11. _____
12. _____
13. _____
14. _____
15. _____
16. _____
17. _____

18. _____

19. _____
20. _____

21. _____
22. _____

23. _____
24. _____
25. _____

You make the call

Using the following illustrations as a guide, describe the process of inspiration and expiration, particularly as it relates to the differences between atmospheric and intrapulmonary pressures.

_____
_____
_____
_____

_____
_____
_____
_____

Atmospheric pressure
(760 mm Hg)

Intrapulmonary pressure
(760 mm Hg)

Intrapleural pressure
(756 mm Hg)

Diaphragm

Atmospheric pressure
(760 mm Hg)

Intrapulmonary pressure
(758 mm Hg)

Intrapleural pressure
(754 mm Hg)

_____
_____
_____

_____
_____
_____

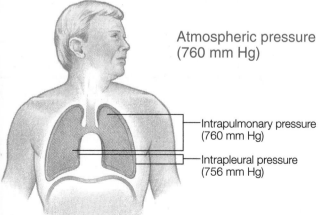

Atmospheric pressure
(760 mm Hg)

Intrapulmonary pressure
(760 mm Hg)

Intrapleural pressure
(756 mm Hg)

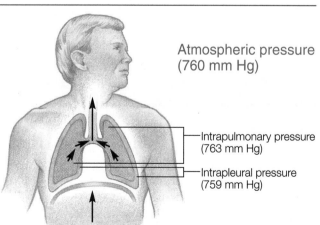

Atmospheric pressure
(760 mm Hg)

Intrapulmonary pressure
(763 mm Hg)

Intrapleural pressure
(759 mm Hg)

Mind sprints

Go the distance by listing as many respiratory changes that occur with aging as you can in 3 minutes.

1. _____
2. _____
3. _____
4. _____
5. _____
6. _____
7. _____
8. _____

First it was facial hair; now my nose is growing. I'm afraid to think what's next.

Power stretch

Unscramble the words on the left to discover what occurs when the body needs increased oxygenation.
Then draw lines linking each process to its appropriate muscular responses.

CODREF RAISINPOINT

_ _ _ _ _ _

_ _ _ _ _ _ _ _ _ _ _ _

VATICE PEANIXTRIO

_ _ _ _ _ _

_ _ _ _ _ _ _ _ _ _

A. Pectoral muscles raise the chest to increase the anteroposterior diameter

B. Internal intercostal muscles contract to shorten the chest's transverse diameter

C. Scalene muscles elevate, fix, and expand the upper chest

D. Posterior trapezius muscles raise the thoracic cage

E. Abdominal rectus muscles pull down the lower chest, depressing the ribs

F. Sternocleidomastoid muscles raise the sternum

Train your brain

Sound out each group of pictures and symbols to reveal information about respiration.

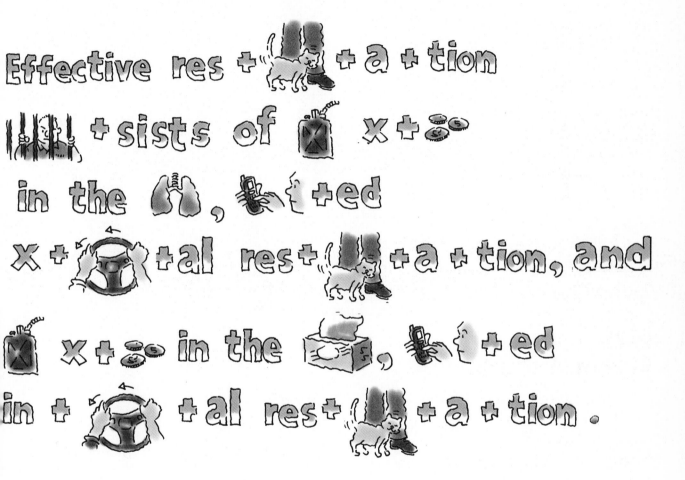

■ Circuit training

Trace the route of pulmonary circulation by drawing arrows between the appropriate boxes.

| Arterioles |

| Right side of heart | | Pulmonary veins | | Gas exchange in lungs |

| Venules |

| Left side of heart | | Right and left pulmonary arteries | | Blood flow to the body |

| Capillary network in alveoli |

Strike out

Some of the following statements about respiration are incorrect. Cross out all of the incorrect statements.

1. Problems within the nervous, musculoskeletal, and pulmonary systems greatly compromise breathing effectiveness.

2. Involuntary breathing results from stimulation of the respiratory center in the medulla and the pons of the brain.

3. The pons controls the rate and depth of respiration.

4. The pons stimulates contraction of the diaphragm and external intercostal muscles, which produce the intrapulmonary pressure changes that cause inspiration.

5. If airflow is disrupted for any reason, airflow distribution follows the path of least resistance.

6. Pulmonary perfusion refers to blood flow from the right side of the heart, through the pulmonary circulation, and into the left side of the heart.

7. Perfusion aids internal respiration.

8. Gravity interferes with perfusion by causing blood to pool in the bases of the lungs.

9. In a healthy lung, ventilation and perfusion should be the same throughout the lung.

10. Gas exchange is most efficient in areas of the lung where ventilation and perfusion are similar.

Mind sprints

Go the distance by seeing if you can list all of the structures within the mediastinum in 2 minutes.

1. _____

2. _____

3. _____

4. _____

5. _____

6. _____

7. _____

8. _____

9. _____

10. _____

■ You make the call

Identify each airflow pattern in the following illustrations. Also describe the conditions that cause each airflow pattern, where in the respiratory passages the pattern usually occurs, and how the pattern affects airway resistance.

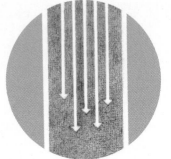      

1. _____   2. _____   3. _____

_____   _____   _____

_____   _____   _____

_____   _____   _____

_____   _____   _____

_____   _____   _____

Coaching session

Factors that increase respiratory workload

These factors can increase respiratory workload, resulting in less efficient ventilation:
- forced breathing
- changes in compliance
- changes in resistance
- low cardiac output
- increased pulmonary and systemic resistance
- abnormal or insufficient hemoglobin.

Match point

Match each of the terms on the left with its definition on the right.

1. Ventilation _____

2. Pulmonary perfusion _____

3. Diffusion _____

4. Metabolic alkalosis _____

5. Metabolic acidosis _____

6. Respiratory alkalosis _____

7. Respiratory acidosis _____

8. Hypoventilation _____

9. Hyperventilation _____

A. Condition resulting from excess bicarbonate retention

B. The distribution of gases into and out of the pulmonary airways

C. Reduced rate and depth of ventilation

D. Process through which oxygen and carbon dioxide molecules move between the alveoli and capillaries

E. Condition resulting from carbon dioxide retention

F. Blood flow from the right side of the heart, through the pulmonary circulation, and into the left side of the heart

G. Increased rate and depth of ventilation

H. Condition resulting from excess acid retention or bicarbonate loss

I. Condition resulting from excessive carbon dioxide elimination

Hit or miss

Some of the following statements about respiration are true; the others are false. Mark each accordingly.

_____ 1. Areas in the lung in which perfusion and ventilation are similar have what's referred to as a ventilation-perfusion match.

_____ 2. Perfusion aids internal respiration.

_____ 3. Both the alveolar epithelium and the capillary endothelium are composed of several layers of elastin and collagen.

_____ 4. Oxygen moves from the alveoli into the bloodstream, where it's taken up by hemoglobin in RBCs.

_____ 5. The presence of carbon dioxide can prevent oxygen from binding with the hemoglobin.

_____ 6. All transported oxygen binds with hemoglobin.

_____ 7. Internal respiration occurs when RBCs release oxygen and absorb carbon dioxide.

_____ 8. In the blood, most carbon dioxide forms bicarbonate.

_____ 9. The lungs control bicarbonate levels by converting bicarbonate to carbon dioxide and water for excretion.

You make the call

The following illustrations show various aspects of ventilation and perfusion. Describe the relationship between ventilation and perfusion shown in each illustration, and identify the physical conditions most likely to cause it.

1. _____

_____

_____

_____

_____

_____

From pulmonary artery

To pulmonary vein

Alveolus

Normal capillary

2. _____

_____

_____

_____

_____

_____

Ventilation blockage

From pulmonary artery

To pulmonary vein

Alveolus

3. _____

_____

_____

_____

_____

_____

From pulmonary artery

To pulmonary vein

Perfusion blockage

Alveolus

Narrowed capillary

4. _____

_____

_____

_____

_____

_____

Ventilation blockage

From pulmonary artery

To pulmonary vein

Perfusion blockage

Alveolus

Key 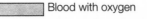 Blood with carbon dioxide ▢ Blood with oxygen  Blood with carbon dioxide and oxygen

11

Gastrointestinal system

Gastrointestinal system review

Gastrointestinal system basics

- Two major components
  - Alimentary canal
  - Accessory organs
- Functions
  - Digestion (breakdown of food and fluid into simple chemicals that can be absorbed into the bloodstream and transported throughout the body)
  - Elimination of waste products (through excretion of stool)

Alimentary canal

- Also called the *GI tract*
- Hollow muscular tube
- Begins in the mouth and extends to the anus
- Includes the pharynx, esophagus, stomach, small intestine, and large intestine

Mouth

- Also called *buccal cavity* or *oral cavity*
- Bounded by the lips, cheeks, palate (roof of the mouth), and tongue and contains the teeth
- Ducts connect the mouth with the three major pairs of salivary glands:
  - Parotid
  - Submandibular
  - Sublingual
- Salivary glands secrete saliva to moisten food during chewing
- Initiates the mechanical breakdown of food

Pharynx

- Cavity that extends from the base of the skull to the esophagus
- Aids swallowing by grasping food and propelling it toward the esophagus

Esophagus

- Muscular tube that extends from the pharynx through the mediastinum to the stomach
- Peristalsis propels liquids and solids through the esophagus into the stomach

- Receives food with relaxation of the cricopharyngeal sphincter (a muscle at the upper border of the esophagus)
- Swallowing triggers the passage of food from the pharynx to the esophagus

Stomach

- Collapsible, pouchlike structure in the left upper part of the abdominal cavity, just below the diaphragm
- Upper border attaches to the lower end of the esophagus
- Lateral surface is called the *greater curvature*
- Medial surface is called the *lesser curvature*
- Size varies with the degree of distention
  - Overeating can cause marked distention, pushing on the diaphragm and causing shortness of breath
- Four main regions:
  - Cardia—lies near the junction of the stomach and esophagus
  - Fundus—enlarged portion above and to the left of the esophageal opening into the stomach
  - Body—middle portion
  - Pylorus—lower portion, which lies near the junction of the stomach and duodenum
- Functions
  - Serves as a temporary storage area for food
  - Begins digestion
  - Breaks down food into chyme (semifluid mass of partially digested food)
  - Moves gastric contents into the small intestine

Small intestine

- Tube that measures about 20′ (6 m) long
- Longest organ of the GI tract
- Three major divisions:
  - Duodenum—longest and most superior division
  - Jejunum—middle portion; shortest segment
  - Ileum—the most inferior portion
- Functions
  - Completing food digestion
  - Absorbing food molecules through its wall into the circulatory system, which then delivers them to body cells
  - Secreting hormones that help control secretion of bile, pancreatic juice, and intestinal juice

INTESTINAL WALL

- Has structural features that significantly increase its absorptive surface area
  - Plicae circulares—circular folds of the intestinal mucosa, or mucous membrane lining.
  - Villi (fingerlike projections on the mucosa) and microvilli (tiny cytoplasmic projections on the surface of epithelial cells)

OTHER STRUCTURES

- Intestinal crypts—simple glands lodged in the grooves separating villi
- Peyer's patches—collections of lymphatic tissue within the submucosa
- Brunner's glands—mucus-secreting glands

Large intestine

- Extends from the ileocecal valve (the valve between the ileum of the small intestine and the first segment of the large intestine) to the anus
- Six segments:
  - Cecum—a saclike structure that makes up the first few inches
  - Ascending colon—rises on the right posterior abdominal wall, then turns sharply under the liver at the hepatic flexure
  - Transverse colon—situated above the small intestine; passing horizontally across the abdomen and below the liver, stomach, and spleen and turning downward at the left colic flexure
  - Descending colon—starts near the spleen and extends down the left side of the abdomen into the pelvic cavity
  - Sigmoid colon—descends through the pelvic cavity, where it becomes the rectum
  - Rectum—last few inches of the large intestine, which terminates at the anus
- Functions
  - Absorbs water
  - Secretes mucus
  - Eliminates digestive wastes

GI tract wall structures

Mucosa

- Also called the *tunica mucosa*
- Innermost layer
- Consists of epithelial and surface cells and loose connective tissue
- Villi—fingerlike projections of the mucosa that secrete gastric and protective juices and absorb nutrients

Submucosa

- Also called the *tunica submucosa*
- Encircles the mucosa

- Composed of:
  - Loose connective tissue
  - Blood vessels
  - Lymphatic vessels
  - Nerve network (submucosal plexus or Meissner's plexus)

Tunica muscularis

- Lies around the submucosa
- Composed of skeletal muscle in the mouth, pharynx, and upper esophagus
- Elsewhere in the tract is made up of longitudinal and circular smooth muscle fibers
  - During peristalsis, longitudinal fibers shorten the lumen length and circular fibers reduce the lumen diameter
  - At points along the tract, circular fibers thicken to form sphincters
  - In the large intestine, these fibers gather into three narrow bands (taeniae coli) down the middle of the colon and pucker the intestine into characteristic pouches (haustra).
- Between the two muscle layers lies another nerve network, the myenteric plexus (also known as *Auerbach's plexus*)
- The stomach wall contains a third muscle layer made up of oblique fibers

Visceral peritoneum

- Outer covering of tract, which covers most of the abdominal organs and lies next to an identical layer, the parietal peritoneum, which lines the abdominal cavity
- Becomes a double-layered fold around the blood vessels, nerves, and lymphatics
- Attaches the jejunum and ileum to the posterior abdominal wall to prevent twisting
- Similar fold attaches the transverse colon to the posterior abdominal wall
- Has many names:
  - In the esophagus and rectum—tunica adventitia
  - Elsewhere in the GI tract—tunica serosa

GI tract innervation

- Distention of the submucosal plexus (Meissner's plexus) stimulates transmission of nerve signals to the smooth muscle, which initiates peristalsis and mixing contractions

Parasympathetic stimulation

- Of the vagus nerve (for most of the intestines) and sacral spinal nerves (for the descending colon and rectum):
  - Increases gut and sphincter tone
  - Increases the frequency, strength, and velocity of smooth muscle contractions as well as motor and secretory activities

Sympathetic stimulation

- By way of the spinal nerves from levels T6 to L2, reduces peristalsis and inhibits GI activity

Accessory organs of digestion

- Contribute hormones, enzymes, and bile, which are vital to digestion

Liver

- Body's largest gland (weighs 3 lb [1.4-kg])
- Highly vascular
- Enclosed in a fibrous capsule in the right upper quadrant of the abdomen
- Mostly covered by the lesser omentum (a fold of the peritoneum), which also anchors it to the lesser curvature of the stomach
  - Hepatic artery and hepatic portal vein as well as the common bile duct and hepatic veins pass through the lesser omentum
- Functions
  - Plays an important role in carbohydrate metabolism
  - Detoxifies various endogenous and exogenous toxins in plasma
  - Synthesizes plasma proteins, nonessential amino acids, and vitamin A
  - Stores essential nutrients, such as vitamins K, D, and B_{12} and iron
  - Removes ammonia from body fluids, converting it to urea for excretion in urine
  - Helps regulate blood glucose levels
  - Secretes bile

LOBES

- Four lobes:
  - Left lobe
  - Right lobe
  - Caudate lobe (behind the right lobe)
  - Quadrate lobe (behind the left lobe).

LOBULES

- Liver's functional unit
- Consists of a plate of hepatic cells, or hepatocytes, that encircle a central vein and radiate outward
- Sinusoids—liver's capillary system
  - Separate the hepatocyte plates from each other
- Reticuloendothelial macrophages (Kupffer's cells)—line the sinusoids and remove bacteria and toxins that have entered the blood through the intestinal capillaries
- Blood flow in lobules:
  - Sinusoids carry oxygenated blood from the hepatic artery and nutrient-rich blood from the portal vein
  - Unoxygenated blood leaves through the central vein and flows through hepatic veins to the inferior vena cava

DUCTS

- Transport bile through the GI tract
- Bile ducts (canaliculi)
  - Allow bile to exit
  - Merge into the right and left hepatic ducts to form the common hepatic duct, which joins the cystic duct from the

gallbladder to form the common bile duct leading to the duodenum

BILE

- Greenish, alkaline liquid composed of water, cholesterol, bile salts, and phospholipids that's continuously secreted by the liver
- Production may increase from stimulation of the vagus nerve, release of the hormone secretin, increased blood flow in the liver, and the presence of fat in the intestine
- Functions
  - Emulsifying (breaking down) fat
  - Promoting intestinal absorption of fatty acids, cholesterol, and other lipids
- Bile salts
  - Aid excretion of lipids from the intestinal tract and absorption of fat-soluble vitamins
  - Liver recycles about 80% into bile, combining them with bile pigments (biliverdin and bilirubin) and cholesterol

Gallbladder

- Pear-shaped organ
- Joined to the ventral surface of the liver by the cystic duct
- Covered by the visceral peritoneum
- Functions
  - Stores and concentrates bile
  - Releases bile into the common bile duct for delivery to the duodenum in response to the contraction and relaxation of Oddi's sphincter

Pancreas

- Somewhat flat organ that lies behind the stomach
  - Head and neck extend into the curve of the duodenum
  - Tail lies against the spleen

EXOCRINE FUNCTIONS

- Secretes more than 1,000 ml of digestive enzymes every day
- Lobules and lobes of the clusters (acini) of enzyme-producing cells release their secretions into ducts that merge into the pancreatic duct

ENDOCRINE FUNCTION

- Involves the islets of Langerhans, located between the acinar cells
- More than 1 million of these islets, which house two cell types:
  - Beta cells secrete insulin to promote carbohydrate metabolism
  - Alpha cells secrete glucagon, a hormone that stimulates glycogenolysis in the liver
  - Both hormones flow directly into the blood
- Hormone release is stimulated by blood glucose levels

PANCREATIC DUCT

- Runs the length of the pancreas
- Joins the bile duct from the gallbladder before entering the duodenum

- Stimulates release of the hormones secretin and cholecystokinin
- Controls the rate and amount of pancreatic secretions

Digestion and elimination

- Digestion starts in the oral cavity, where chewing (mastication), salivation (the beginning of starch digestion), and swallowing (deglutition) occur
- When a person swallows, the hypopharyngeal sphincter in the upper esophagus relaxes, allowing food to enter the esophagus
- In the esophagus, the glossopharyngeal nerve activates peristalsis, which moves the food down toward the stomach
- As food passes through the esophagus, glands in the esophageal mucosal layer secrete mucus, which lubricates the bolus and protects the mucosal membrane from damage caused by poorly chewed foods

Cephalic phase of digestion

- Begins by the time the bolus travels toward the stomach
- Stomach secretes digestive juices (hydrochloric acid and pepsin)

Gastric phase of digestion

- Initiated by stomach wall stretching caused by food entering the stomach through the cardiac sphincter
- Distention of the stomach wall stimulates the stomach to release gastrin

GASTRIN

- Digestive secretion that consists mainly of pepsin, hydrochloric acid, intrinsic factor, and proteolytic enzymes
- Stimulates the stomach's motor functions and secretion of gastric juice by the gastric glands
- Highly acidic (pH of 0.9 to 1.5)

Intestinal phase of digestion

- Characterized by little absorption (except for alcohol)
- Peristaltic contractions churn food into tiny particles and mix it with gastric juices, forming chyme
- Stronger peristaltic waves move the chyme into the antrum
- In the atrium, chyme backs up against the pyloric sphincter before being released into the duodenum
- Release into the duodenum triggers the intestinal phase of digestion

STOMACH EMPTYING

- Rate depends on several factors, including:
  - Gastrin release
  - Neural signals generated when the stomach wall distends
  - Enterogastric reflex
- During enterogastric reflex , the duodenum releases secretin and gastric-inhibiting peptide and the jejunum secretes cholecystokinin—all of which decrease gastric motility
- Small intestine performs most of the work of digestion and absorption

- In the small intestine, intestinal contractions and various digestive secretions break down carbohydrates, proteins, and fats
- Intestinal mucosa absorbs carbohydrates, proteins, and fats into the bloodstream (along with water and electrolytes)
- Nutrients are then available for use by the body.
- By the time chyme passes through the small intestine and enters the ascending colon of the large intestine, it has been reduced to mostly indigestible substances
- Food bolus begins its journey through the large intestine, where the ileum and cecum join with the ileocecal pouch
- Bolus moves up the ascending colon, past the right abdominal cavity to the liver's lower border
- Bolus crosses horizontally below the liver and stomach by way of the transverse colon
- Bolus descends the left abdominal cavity to the iliac fossa through the descending colon
- Bolus travels through the sigmoid colon to the lower midline of the abdominal cavity, then to the rectum, and finally to the anal canal
- Anus opens to the exterior through two sphincters
  - Internal sphincter contains thick, circular, smooth muscle under autonomic control
  - External sphincter contains skeletal muscle under voluntary control

ROLE IN ABSORPTION

- Large intestine produces no hormones or digestive enzymes
- Large intestine continues the absorptive process
- Proximal half of the large intestine absorbs all but about 100 ml of the remaining water in the colon as well as large amounts of both sodium and chloride

BACTERIAL ACTION

- Large intestine harbors the bacteria *Escherichia coli*, *Enterobacter aerogenes*, *Clostridium perfringens*, and *Lactobacillus bifidus*
  - Help synthesize vitamin K and break down cellulose into a usable carbohydrate
  - Also produces flatus, which helps propel stool toward the rectum
- Mucosa of the large intestine produces alkaline secretions from tubular glands composed of goblet cells
- Alkaline mucus lubricates the intestinal walls as food pushes through, protecting the mucosa from acidic bacterial action
- In the lower colon, long and relatively sluggish contractions cause propulsive waves, or mass movements
- Occurring several times per day, these movements propel intestinal contents into the rectum and produce the urge to defecate
- Defecation normally results from the defecation reflex (a sensory and parasympathetic nerve-mediated response) along with voluntary relaxation of the external anal sphincter

■ Cross-training

Here's an exercise to help stretch your knowledge of terms related to the GI system.

Across

2. Longest and most inferior portion of the small intestine

5. Shortest and most superior division of the small intestine

7. Saclike structure comprising the first few inches of the large intestine

9. Term referring to swallowing

10. Cavity extending from the base of the skull to the esophagus

13. Collapsible, pouchlike structure in the left upper portion of the abdominal cavity

14. Fingerlike projections on the mucosa of the intestinal wall

15. Serve as the liver's capillary system

16. Muscular tube extending from the pharynx to the stomach

Down

1. The last few inches of the large intestine, terminating at the anus

3. The stomach breaks down food into this semifluid substance

4. Middle portion of the small intestine

6. Term referring to chewing

8. Term for hepatic cells

11. Innermost layer of the GI tract

12. Name for the liver's functional unit

■■
■ Finish line

Label the structures of the GI system shown in the following illustration.

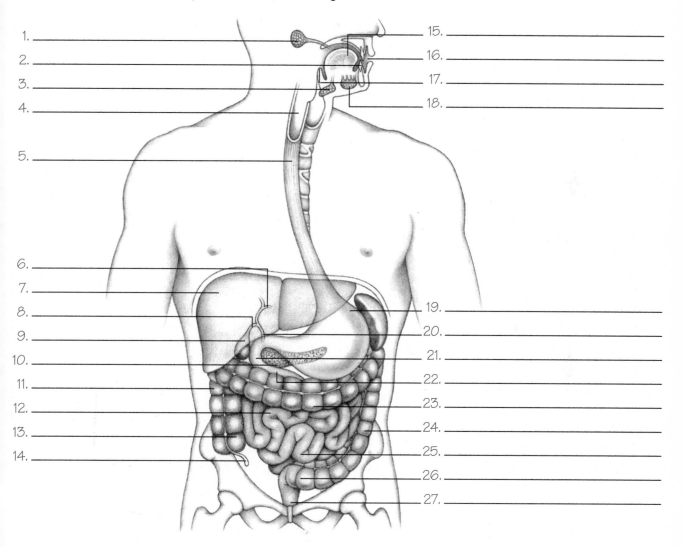

1. _____
2. _____
3. _____
4. _____
5. _____
6. _____
7. _____
8. _____
9. _____
10. _____
11. _____
12. _____
13. _____
14. _____

15. _____
16. _____
17. _____
18. _____
19. _____
20. _____
21. _____
22. _____
23. _____
24. _____
25. _____
26. _____
27. _____

Coaching session
Salivary glands

Three pairs of salivary glands secrete saliva to moisten food during chewing:
- parotid
- submandibular
- sublingual.

■ Match point

Match each of the structures on the left with its function on the right.

1. Mouth _____
2. Pharynx _____
3. Esophagus _____
4. Salivary glands _____
5. Villi and microvilli _____

A. Secrete saliva to moisten food during chewing
B. Aids swallowing by grasping food and propelling it toward the esophagus
C. Initiates the mechanical breakdown of food
D. Conducts food between the pharynx to the stomach
E. Increase the absorptive area of the intestinal wall

Overeating can markedly distend the stomach. This pushes on the diaphragm and causes shortness of breath.

Power stretch

Unscramble the following words to discover some of the structures of the GI tract. Then draw lines to connect each structure with its particular functions.

MATCHSO

_ _ _ _ _ _ _

MALLS TENNISTIE

_ _ _ _ _

_ _ _ _ _ _ _ _ _

GLARE NINESETIT

_ _ _ _ _

_ _ _ _ _ _ _ _ _

A. Absorbs water

B. Moves gastric contents into the small intestine

C. Absorbs food molecules through its wall into the circulatory system

D. Serves as a temporary storage area for food

E. Eliminates digestive wastes

F. Completes food digestion

G. Begins digestion

H. Secretes mucus

I. Breaks down food into chyme

J. Secretes hormones that help control secretion of bile, pancreatic fluid, and intestinal fluid

Train your brain

Sound out each group of pictures and symbols to reveal information related to digestion.

Match point

Match each of the stomach regions listed below with its location.

1. Cardia _____
2. Fundus _____
3. Body _____
4. Pylorus _____

A. Middle portion of the stomach

B. Portion near the junction of the stomach and esophagus

C. Enlarged portion above and to the left of the esophageal opening into the stomach

D. Lower portion of the stomach, near the junction of the stomach and duodenum

Train your brain

Sound out each group of pictures and symbols to reveal information related to physiology of the stomach.

1.

2.

Finish line

Label the structures of the oral cavity shown in this illustration.

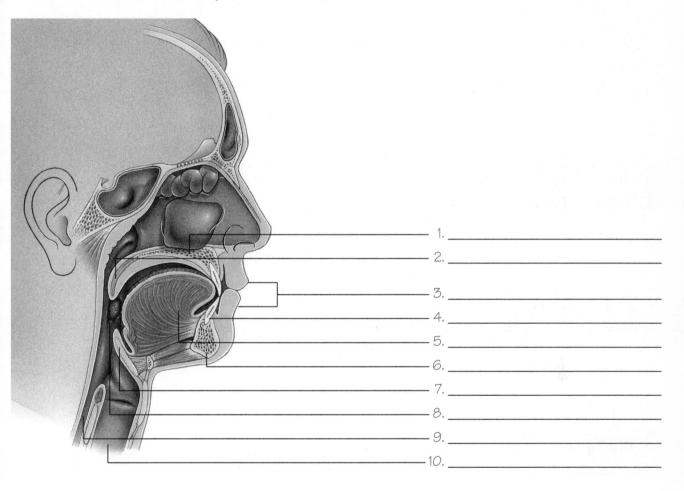

1. _____
2. _____
3. _____
4. _____
5. _____
6. _____
7. _____
8. _____
9. _____
10. _____

Pep talk

> Learning without thought is labor lost; thought without learning is perilous.
> —Confucius

■ Power stretch

Unscramble the following words to discover the names of the layers of the GI tract wall. Then draw lines to connect each layer with its particular characteristics.

CAMSOU

— — — — — —

SCUBASUMO

— — — — — — — — —

ACTINU ACURLISMUS

— — — — — —

— — — — — — — — — —

A. The innermost layer

B. Lies around the submucosa

C. Also called the *tunica submucosa*

D. Also called the *tunica mucosa*

E. Consists of epithelial and surface cells and loose connective tissue

F. Composed of loose connective tissue, blood and lymphatic vessels, and a nerve network

G. Contains fingerlike projections called villi

H. Encircles the mucosa

I. Composed of skeletal muscle in the mouth, pharynx, and upper esophagus; elsewhere is made up of longitudinal and circular smooth muscle fibers

■ Match point

Match each of the terms below with its definition.

1. Peyer's patches _____

2. Plicae circulares _____

3. Submucosal plexus (also called Meissner's plexus) _____

4. Taeniae coli _____

5. Myenteric plexus (also called Auerbach's plexus) _____

6. Intestinal crypts _____

A. Narrow bands of fibers down the middle of the colon

B. Simple glands lodged in the grooves separating villi

C. Circular folds of the intestinal mucosa

D. Collections of lymphatic tissue within the submucosa of the small intestine

E. Nerve network in the submucosa

F. Nerve network lying between the two muscle layers of the large intestine

Strike out

Some of the following statements about the GI system are incorrect. Cross out all of the incorrect statements.

1. Distention of the submucosal plexus stimulates transmission of nerve signals to the GI tract to slow peristalsis.

2. Parasympathetic stimulation of the vagus and sacral spinal nerves increases gut and sphincter tone.

3. Parasympathetic stimulation of the GI tract relaxes the smooth muscle of the GI tract.

4. Sympathetic stimulation reduces peristalsis and inhibits GI activity.

5. The visceral peritoneum lines the abdominal cavity.

6. The visceral peritoneum attaches the jejunum and ileum to the posterior abdominal wall.

Hit or miss

Some of the following statements are true; the others are false. Mark each accordingly.

_____ 1. The large intestine is the longest organ of the GI tract.

_____ 2. The liver is the body's largest gland.

_____ 3. The lesser omentum is the lateral curve of the liver.

_____ 4. The liver consists of three lobes.

_____ 5. The liver's functional unit is the hepatocyte.

_____ 6. Sinusoids are the liver's capillary system.

■ Mind sprints

Go the distance by listing as many of the liver's functions as you can in 1 minute.

1. _____
2. _____
3. _____
4. _____
5. _____
6. _____
7. _____

Time for a little speed work in this mental workout. On your mark, get set, go!

■ Match point

Match each the following GI structures to its function.

1. Gallbladder _____
2. Pancreas _____
3. Common bile duct _____
4. Canaliculi _____
5. Pancreatic duct _____

A. Transports bile to the duodenum

B. Joins the bile duct from the gallbladder before entering the duodenum

C. Stores and concentrates bile

D. Performs both endocrine and exocrine functions

E. Merges into the right and left hepatic ducts to form the common hepatic duct

■ Finish line

Label the structures of the biliary tract shown in this illustration.

1. _____

2. _____

3. _____

4. _____

5. _____

6. _____

7. _____

8. _____

9. _____

10. _____

11. _____

12. _____

13. _____

14. _____

15. _____

■ Match point

Match each of the following GI hormones and enzymes to its function.

1. Gastrin _____

2. Gastric inhibitory peptides _____

3. Secretin _____

4. Cholecystokinin _____

5. Hydrochloric acid _____

6. Pepsinogen _____

7. Intrinsic factor _____

8. Insulin _____

9. Glucagon _____

A. Breaks proteins down into polypeptides

B. Stimulates gallbladder contraction and secretion of pancreatic fluid

C. Stimulates gastric secretion and motility

D. Promotes carbohydrate metabolism

E. Inhibits gastric secretion and motility

F. Promotes vitamin B_{12} absorption in the small intestine

G. Stimulates secretion of bile and alkaline pancreatic fluid

H. Degrades pepsinogen, maintains acid environment, and inhibits excess bacterial growth

I. Stimulates glycogenolysis in the liver

You make the call

Describe the three parts of the neural pattern that initiates swallowing, using the illustration below as a guide.

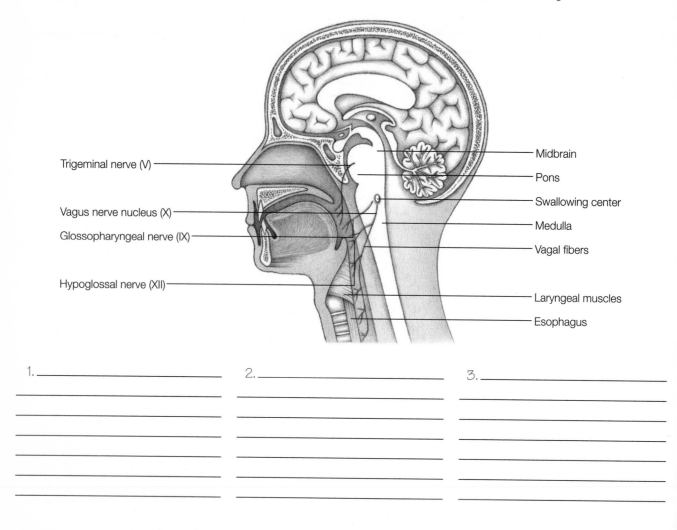

1.

2.

3.

Hit or miss

Some of the following statements are true; the others are false. Mark each accordingly.

_____ 1. Nearly all digestion and absorption takes place in the small intestine.

_____ 2. Multiple projections on the intestinal mucosa help facilitate the passage of food through the intestinal tract.

_____ 3. Brunner's glands secrete large amounts of mucus to lubricate and protect the duodenum.

_____ 4. Duodenal argentaffin cells produce the hormones secretin and cholecystokinin.

_____ 5. The intestinal phase of digestion begins when food enters the large intestine.

Starting lineup

Test your knowledge of digestion and elimination. Put the following steps in the order they occur.

Gastrin is released.
Excretion exits through the anal canal.
Food bolus enters the esophagus and is lubricated with mucus.
Intestinal contractions and various digestive secretions break down carbohydrates, proteins, and fats.
The stomach stretches.
Chewing, salivation, and swallowing occur.
Gastric juices mix with and break down the food, forming chyme.
Chyme moves into the antrum and then on to the duodenum.
Glossopharyngeal nerve activates peristalsis, moving the food toward the stomach.
Bolus passes through the large intestine and on to the sigmoid colon.

1.

2.

3.

4.

5.

6.

7.

8.

9.

10.

Mind sprints

Go the distance by listing within 30 seconds the three factors that influence the rate of stomach emptying.

1. _____

2. _____

3. _____

Hit or miss

Some of the following statements are true; the others are false. Mark each accordingly.

_____ 1. The anus opens to the exterior through two sphincters.

_____ 2. Both anal sphincters are under voluntary control.

_____ 3. The large intestine produces various hormones and enzymes to continue to digestive process.

_____ 4. The proximal half of the large intestine absorbs water as well as large amounts of sodium and chloride.

_____ 5. Under normal circumstances, the large intestine should be free of bacteria.

Train your brain

Sound out each group of pictures and symbols to reveal information related to digestion.

Nutrition and metabolism

Warm-up

Nutrition and metabolism review

Nutrition

- Taking in, assimilating, and utilizing nutrients
- Nutrients in foods must be broken down into components
- Within cells, the products of digestion undergo further chemical reactions
- *Metabolism* refers to the sum of these chemical reactions
- Food substances are transformed into energy or materials that the body can use or store
- Metabolism involves two processes:
  - Anabolism—synthesis of simple substances into complex ones
  - Catabolism—breakdown of complex substances into simpler ones or into energy
- The body needs a continual supply of water and various nutrients for growth and repair
- Virtually all nutrients come from digested food
- Three major types of nutrients required by the body:
  - Carbohydrates
  - Proteins
  - Lipids
- Vitamins—essential for normal metabolism and contribute to the enzyme reactions that promote the metabolism of carbohydrates, proteins, and lipids
- Minerals—participate in such essential functions as enzyme metabolism and membrane transfer of essential elements

Carbohydrates

- Organic compounds composed of carbon, hydrogen, and oxygen
- Yield 4 kcal/g when used for energy
- Include sugars, which function as the body's primary energy source

Monosaccharides

- Simple sugars that can't be split into smaller units by hydrolysis
- Subdivided into polyhydroxy aldehydes or ketones based on whether the molecule consists of an aldehyde group or a ketone group
- An aldehyde contains the characteristic group CHO
- Polyhydroxy—linking of carbon atoms to a hydroxyl (OH) group
- Ketone—contains the carbonyl group CO and carbon groups attached to the carbonyl carbon

Disaccharides

- Synthesized from monosaccharides
- Consist of two monosaccharides minus a water molecule
- Examples:
  - Sucrose: a combination of a glucose molecule and a fructose molecule
  - Lactose: sugar in milk; a combination of a glucose molecule and a galactose molecule
  - Maltose: a combination of two glucose molecules

Polysaccharides

- Synthesized from monosaccharides
- Consist of a long chain (polymer) of more than 10 monosaccharides linked by glycoside bonds
- Example: Glycogen (the body builds glycogen by using excess sugars [monosaccharides] and stores it for future use)
- Also ingested and broken down into simple sugars used for fuel
- Fiber—a polysaccharide that can't be broken down into simple sugars and, therefore, can't be used by the body for energy

Proteins

- Complex nitrogenous organic compounds containing amino acid chains and, some, sulfur and phosphorus
- Used mainly for growth and repair of body tissues
- When used for energy, yield 4 kcal/g
- Some combine with lipids to form lipoproteins or with carbohydrates to form glycoproteins
- Each is synthesized on a ribosome as a straight chain

Amino acids

- Building blocks of proteins
- Each contains a carbon atom to which a carboxyl (COOH) group and an amino group are attached
  - Unite by condensation of the COOH group on one amino acid with the amino group of the adjacent amino acid
  - Reaction releases a water molecule and creates a linkage called a *peptide bond*
- Sequence and types in a chain determine the nature of the protein
  - Chemical attractions cause the chain to coil or twist into a specific shape
  - Shape determines function

Lipids

- Organic compounds that don't dissolve in water but do dissolve in alcohol and other organic solvents

- Concentrated form of fuel that yields approximately 9 kcal/g when used for energy

Fats

- Also referred to as *triglycerides*
- Most common lipids
- Contain three molecules of fatty acid combined with one molecule of glycerol
  - Fatty acid—long-chain compound with an even number of carbon atoms and a terminal COOH group

Phospholipids

- Complex lipids that are similar to fat
- Have a phosphorus- and nitrogen-containing compound that replaces one of the fatty acid molecules
- Major structural components of cell membranes

Steroids

- Complex molecules in which carbon atoms form four cyclic structures attached to various side chains
- Contain no glycerol or fatty acid molecules
- Examples: cholesterol, bile salts, and sex hormones

Vitamins and minerals

Vitamins

- Organic compounds that are needed in small quantities for normal metabolism, growth, and development
- Classified as water-soluble or fat-soluble
  - Examples of water-soluble vitamins: B complex vitamins and C vitamins
  - Examples of fat-soluble vitamins: vitamins A, D, E, and K

Minerals

- Inorganic substances that play important roles in:
  - enzyme metabolism
  - membrane transfer of essential compounds
  - regulation of acid-base balance
  - osmotic pressure
  - muscle contractility
  - nerve impulse transmission
  - growth
- Found in bones, hemoglobin, thyroxine, and vitamin B_1
- Classified as major (more than 0.005% of body weight) or trace (less than 0.005% of body weight)
  - Major minerals include calcium, chloride, magnesium, phosphorus, potassium, sodium
  - Trace minerals include chromium, cobalt, copper, fluorine, iodine, iron, manganese, molybdenum, selenium, zinc

Absorption and digestion

- Nutrients must be digested in the GI tract by enzymes that split large units into smaller ones
  - In hydrolysis, a compound unites with water and then splits into simpler compounds

- The smaller units are then absorbed from the small intestine and transported to the liver through the portal venous system

Carbohydrate digestion and absorption

- Enzymes break down complex carbohydrates
- In the oral cavity, salivary amylase initiates starch hydrolysis into disaccharides
- In the small intestine, pancreatic amylase continues this process
- Disaccharides in the intestinal mucosa hydrolyze disaccharides into monosaccharides
- Lactase splits lactose into glucose and galactose
- Sucrase hydrolyzes the compound sucrose into glucose and fructose
- Monosaccharides, such as glucose, fructose, and galactose, are absorbed through the intestinal mucosa
- Monosaccharides are then transported through the portal venous system to the liver
- In the liver, enzymes convert fructose and galactose to glucose
- Ribonucleases and deoxyribonucleases break down nucleotides from deoxyribonucleic acid (DNA) and ribonucleic acid (RNA) into pentoses and nitrogen-containing compounds (nitrogen bases)
- Like glucose, these compounds are absorbed through the intestinal mucosa

Protein digestion and absorption

- Enzymes digest proteins by hydrolyzing the peptide bonds that link the amino acids of the protein chains, which restores water molecules
- Gastric pepsin breaks proteins into:
  - polypeptides
  - pancreatic trypsin
  - chymotrypsin
  - carboxypeptidase, which converts polypeptides to peptides
- Intestinal mucosal peptidases break down peptides into their constituent amino acids
- After being absorbed through the intestinal mucosa by active transport mechanisms, these amino acids travel through the portal venous system to the liver
- The liver converts the amino acids not needed for protein synthesis into glucose

Lipid digestion and absorption

- Pancreatic lipase breaks down fats and phospholipids into a mixture of:
  - glycerol
  - short- and long-chain fatty acids
  - monoglycerides
- The portal venous system then carries these substances to the liver
- Lipase hydrolyzes the bonds between glycerol and fatty acids, a process that restores the water molecules released when the bonds were formed

- Glycerol diffuses directly through the mucosa
- Short-chain fatty acids diffuse into the intestinal epithelial cells and are carried to the liver via the portal venous system
- Long-chain fatty acids and monoglycerides in the intestine dissolve in the bile salt micelles and then diffuse into the intestinal epithelial cells
- Lipase breaks down absorbed monoglycerides into glycerol and fatty acids
- In the smooth endoplasmic reticulum of the epithelial cells, fatty acids and glycerol recombine to form fats

Chylomicrons

- Lipoprotein particles formed by a small amount of cholesterol and phospholipids that are coated with triglycerides
- Collect in the intestinal lacteals (lymphatic vessels) and are carried through lymphatic channels
- After entering the circulation through the thoracic duct, distributed to body cells
- In the cells, undergo extraction of fats, which are broken down by enzymes into fatty acids and glycerol
- Then absorb into and recombine in fat cells, reforming triglycerides for storage and later use

Carbohydrate metabolism

- Preferred energy fuel of human cells
- Most in absorbed food are quickly catabolized for the release of energy

Glucose to energy

- All ingested carbohydrates are converted to glucose, the body's main energy source
- Glucose not needed for immediate energy is stored as glycogen or converted to lipids
- Energy from glucose catabolism is generated in three phases
  - Glycolysis
  - Krebs cycle (also called the *citric acid cycle*)
  - Electron transport system

GLYCOLYSIS

- Process by which enzymes break down the 6-carbon glucose molecule into two 3-carbon molecules of pyruvic acid (pyruvate)
- Occurs in the cell cytoplasm
- Doesn't use oxygen
- Glycolysis yields energy in the form of adenosine triphosphate (ATP).
- Pyruvic acid releases a carbon dioxide molecule, which is converted in the mitochondria to a two-carbon acetyl fragment, which combines with coenzyme A (CoA) (a complex organic compound) to form acetyl CoA

KREBS CYCLE

- The second phase in glucose catabolism
- Occurs in the mitochondria
- Requires oxygen

- Pathway by which a molecule of acetyl CoA is oxidized by enzymes to yield energy
  - Two-carbon acetyl fragments of acetyl CoA enter the Krebs cycle by joining to the four-carbon compound oxaloacetic acid to form citric acid, a six-carbon compound
  - CoA molecule detaches from the acetyl group, becoming available to form more acetyl CoA molecules
  - Enzymes convert citric acid into intermediate compounds and eventually convert it back into oxaloacetic acid
  - Cycle begins again
- In addition to liberating carbon dioxide and generating energy, each cycle releases hydrogen atoms, which are picked up by the coenzymes nicotinamide adenine dinucleotide (NAD) and flavin adenine dinucleotide (FAD)

ELECTRON TRANSPORT SYSTEM

- Last phase of carbohydrate catabolism
- Occurs in the mitochondria
- Requires oxygen
- In this phase
  - Carrier molecules on the inner mitochondrial membrane pick up electrons from the hydrogen atoms carried by NAD and FAD; each hydrogen atom contains a hydrogen ion and an electron
  - Carrier molecules transport the electrons through a series of enzyme-catalyzed oxidation-reduction reactions in the mitochondria
  - Oxygen plays a crucial role by attracting electrons along the chain of carriers in the transport system
  - During oxidation, a chemical compound loses electrons; during reduction, it gains electrons
  - These reactions release the energy contained in the electrons and generate ATP
- After passing through the electron transport system, the hydrogen ions produced in the Krebs cycle combine with oxygen to form water

Liver and muscle cells

- Help regulate blood glucose levels

Liver

- When glucose levels exceed the body's immediate needs, hormones stimulate the liver to convert glucose into glycogen or lipids
  - Glycogen forms through glycogenesis
  - Lipids form through lipogenesis
- When blood glucose levels drop excessively, the liver can form glucose by two processes:
  - Breakdown of glycogen to glucose through glycogenolysis
  - Synthesis of glucose from amino acids through gluconeogenesis

Muscle cells

- Convert glucose to glycogen for storage but lack the enzymes to convert glycogen back to glucose when needed

- During vigorous muscular activity, when oxygen requirements exceed the oxygen supply, break down glycogen to yield lactic acid and energy
  - Lactic acid then builds up in the muscles, and muscle glycogen is depleted
  - Some of the lactic acid diffuses from muscle cells, is transported to the liver, and is reconverted to glycogen
  - The liver converts the newly formed glycogen to glucose, which travels through the bloodstream to the muscles and reforms into glycogen
  - When muscle exertion stops, some of the accumulated lactic acid converts back to pyruvic acid
  - Pyruvic acid is oxidized completely to yield energy by means of the Krebs cycle and electron transport system

Protein metabolism

- Absorbed as amino acids and carried by the portal venous system to the liver and then throughout the body by blood
- When absorbed, mix with other amino acids in the body's amino acid pool

Amino acid conversion

- Because the body can't store amino acids, they are converted to protein or glucose or catabolized to provide energy
- Amino acids must be transformed by deamination or transamination

DEAMINATION

- An amino group ($-NH_2$) splits off from an amino acid molecule to form a molecule of ammonia and one of keto acid
- Most of the ammonia is converted to urea and excreted in the urine

TRANSAMINATION

- An amino group is exchanged for a keto group in a keto acid through the action of transaminase enzymes
- During this process, the amino acid is converted to a keto acid and the original keto acid is converted to an amino acid

Amino acid synthesis

- Proteins are synthesized from 20 amino acids from the body's amino acid pool
- Amino acids not used for protein synthesis can be converted to keto acids and metabolized by the Krebs cycle and the electron transport system to produce energy
- Amino acids can also be converted to other nutrients such as fats
- Amino acids not used for protein synthesis may be converted to pyruvic acid and then to acetyl CoA
  - Acetyl CoA fragments condense to form long-chain fatty acids, a process that's the reverse of fatty acid breakdown
  - These fatty acids then combine with glycerol to form fats
  - Amino acids are converted to pyruvic acid, which may then be converted to glucose

Lipid metabolism

- Lipids are stored in adipose tissue within cells until required for use as fuel
- When needed for energy, each fat molecule is hydrolyzed to glycerol and three molecules of fatty acid
- Glycerol can be converted to pyruvic acid and then to acetyl CoA, which enters the Krebs cycle

Ketone body formation

- The liver normally forms ketone bodies from acetyl CoA fragments, derived largely from fatty acid catabolism
- Acetyl CoA molecules yield three types of ketone bodies:
  - Acetoacetic acid
  - Beta-hydroxybutyric acid
  - Acetone
- Acetoacetic acid results from the combination of two acetyl CoA molecules and subsequent release of CoA from these molecules
- Beta-hydroxybutyric acid forms when hydrogen is added to the oxygen atom in the acetoacetic acid molecule
- The term *beta* indicates the location of the carbon atom containing the OH group
- Acetone forms when the COOH group of acetoacetic acid releases carbon dioxide
- Muscle tissue, brain tissue, and other tissues oxidize these ketone bodies for energy

EXCESSIVE KETONE FORMATION

- Under certain conditions, the body produces more ketone bodies than it can oxidize for energy
- The body must then use fat instead of glucose as its primary energy source
- Use of fat instead of glucose for energy leads to an excess of ketone bodies

LIPID FORMATION

- Excess amino acids can be converted to fat through keto acid–acetyl CoA conversion
- Glucose may be converted to pyruvic acid and then to acetyl CoA
- CoA is converted into fatty acids and then fat

Hormonal regulation of metabolism

- Normal body functions necessitate that blood glucose levels stay within a certain range
- Hormones regulate blood glucose level by stimulating the metabolic processes that restore a normal level in response to blood glucose changes
- Insulin, produced by the pancreatic islet cells, is the only hormone that significantly reduces blood glucose level
  - Insulin promotes cell uptake and use of glucose as an energy source
  - Insulin promotes glucose storage as glycogen (glycogenesis) and lipids (lipogenesis)

■ Batter's box

Fill in the blanks with the appropriate words. *Hint:* Words may be used more than once.

Nutritious facts

_____ refers to the intake, assimilation, and utilization of nutrients. The
$\underset{1}{}$

crucial nutrients in foods must be broken down into _____. Within cells, the
$\underset{2}{}$

products of digestion undergo further _____.
$\underset{3}{}$

Breaking it down

Metabolism refers to the sum of these _____. Through metabolism, food
$\underset{4}{}$

substances are transformed into _____ or materials that the body can use or
$\underset{5}{}$

store. Metabolism involves two processes: _____ (the synthesis of simple
$\underset{6}{}$

substances into complex ones) and _____ (the breakdown of complex
$\underset{7}{}$

substances into simpler ones or into energy).

Nutrient-dense data

The body needs a continual supply of _____ and various _____
$\underset{8}{}$ $\underset{9}{}$

for growth and repair. The three major types of nutrients required by the body are

_____, _____ , and _____. _____
$\underset{10}{}$ $\underset{11}{}$ $\underset{12}{}$ $\underset{13}{}$

and _____ are also essential for normal metabolism.
$\underset{14}{}$

Options

anabolism

catabolism

carbohydrates

chemical
  reactions

components

energy

lipids

minerals

nutrients

nutrition

proteins

vitamins

water

Cross-training

Here's an exercise to help stretch your knowledge of terms related to nutrition and metabolism.

Across

3. Sugar used in brewing and distilling

5. Simple sugars that can't be splint into smaller units

8. Hormone that reduces blood glucose levels

9. Enzyme that breaks down complex carbohydrates

10. The body's main energy source

Down

1. Organic compounds that are classified as water-soluble or fat-soluble

2. Sugar found in milk

4. Complex nitrogenous organic compounds containing amino acid chains

6. Process by which a compound unites with water and then splits into simpler compounds

7. Common table sugar

Power stretch

Unscramble the words on the left to discover the names of three major types of nutrients. Then draw lines to link each nutrient to the correct information. Note that some information may apply to more than one nutrient.

OPENSTIR

_ _ _ _ _ _ _ _

SLIDPI

_ _ _ _ _ _

ACABREDHOSTRY

_ _ _ _ _ _ _ _ _ _ _ _ _

A. Concentrated form of fuel

B. Used mainly for growth and repair of body tissues

C. Yield 9 kcal/g when used for energy

D. Organic compounds composed of carbon, hydrogen, and oxygen

E. Complex nitrogenous organic compounds containing amino acid chains

F. Yield 4 kcal/g when used for energy

G. Organic compounds that don't dissolve in water but do dissolve in alcohol and other organic solvents

H. The body's primary energy source

I. Composed of amino acids

J. Classified as monosaccharides, disaccharides, and polysaccharides

K. Major types include fats, phospholipids, and steroids

Memory jogger

Remember the three P's of proteins:

1. P_____ = 2 to 10 amino acids

2. P_____ = 10 or more amino acids

3. P_____ = 50 or more amino acids.

■ Match point

Match each vitamin listed below with its functions.

1. Vitamin C _____
2. Vitamin B$_1$ _____
3. Vitamin B$_2$ _____
4. Vitamin B$_6$ _____
5. Folic acid _____
6. Niacin _____
7. Vitamin B$_{12}$ _____
8. Vitamin A _____
9. Vitamin D _____
10. Vitamin E _____
11. Vitamin K _____

A. Body tissue repair and maintenance, night vision
B. Blood formation, carbohydrate metabolism
C. Collagen production, healing, infection resistance
D. Antibody formation, digestion, DNA and RNA synthesis
E. Calcium and phosphorus metabolism (bone formation)
F. Circulation, cholesterol level reduction
G. RBC formation, energy metabolism, and cell respiration
H. Liver synthesis of prothrombin and other blood-clotting factors
I. Cell growth and reproduction, digestion, liver function
J. RBC formation, tissue growth, nerve cell maintenance, appetite stimulation
K. Aging retardation, anticlotting factor, diuresis, myocardial perfusion

You seem to have absorbed a lot of information about nutrition. This game on vitamins should be a slam dunk.

Hit or miss

Some of the following statements are true; the others are false. Mark each accordingly.

_____ 1. Enzymes begin breaking down complex carbohydrates in the mouth.

_____ 2. Monosaccharides (such as glucose, fructose, and galactose) are absorbed through the intestinal mucosa and transported to the liver.

_____ 3. Enzymes digest proteins by hydrolyzing the peptide bonds between the amino acids.

_____ 4. The liver converts amino acids not needed for protein synthesis into fat.

_____ 5. Most fat digestion occurs in the stomach.

_____ 6. Triglycerides are coated with a thin layer of protein to form chylomicrons.

_____ 7. Carbohydrates require a prolonged digestion process before they can be used for energy.

Train your brain

Sound out each group of pictures and symbols to reveal information related to metabolism.

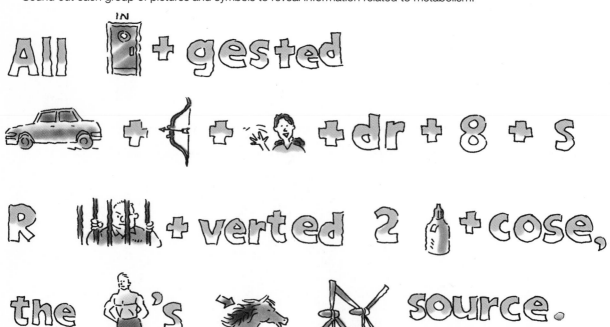

■ You make the call

Describe the processes of glycolysis and Krebs cycle, using the flowchart shown here as a guideline.

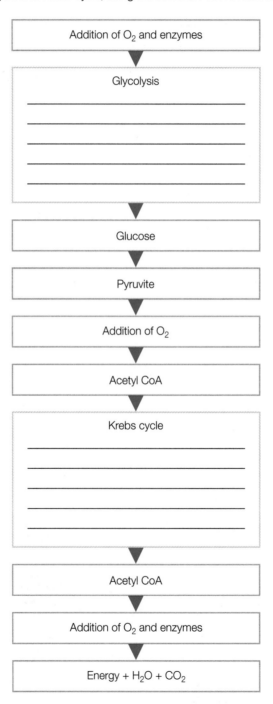

Addition of O_2 and enzymes

Glycolysis
_____
_____
_____
_____
_____

Glucose

Pyruvite

Addition of O_2

Acetyl CoA

Krebs cycle
_____
_____
_____
_____
_____

Acetyl CoA

Addition of O_2 and enzymes

Energy + H_2O + CO_2

Strike out

Some of the following statements about amino acids are incorrect. Cross out all of the incorrect answers.

1. Proteins are absorbed as amino acids and carried to the liver and then throughout the body.

2. Amino acids are the structural units of proteins.

3. The body stores excess amino acids in the liver.

4. Amino acids are classified as essential or nonessential based on whether they're needed for protein synthesis.

5. Amino acids can be converted to other nutrients, such as fats or glucose.

6. Use of fat instead of glucose for energy leads to an amino acid deficit.

Train your brain

Sound out each group of pictures and symbols to reveal information related to metabolism.

13

Urinary system

Warm-up

Urinary system review

Urinary system basics

- Removes wastes from the body
- Helps govern acid-base balance by retaining or excreting hydrogen ions
- Regulates fluid and electrolyte balance

Structures

Kidneys

- Bean-shaped, highly vascular organs
- Renal cortex (outer region)—contains blood-filtering mechanisms
- Renal medulla (middle region)—contains 8 to 12 renal pyramids and calyces
- Protected in front by the abdomen and behind by the muscles attached to the vertebral column
- Each has an adrenal gland on top, which influences blood pressure and the kidney's retention of sodium and water
- Highly vascular, receiving about 20% of the blood pumped by the heart each minute
- Functions:
  - Elimination of wastes and excess ions (as urine)
  - Blood filtration (by regulating chemical composition and volume of blood)
  - Maintenance of fluid-electrolyte and acid-base balances
  - Production of erythropoietin (a hormone that stimulates red blood cell [RBC] production) and enzymes (such as renin, which governs blood pressure and kidney function)
  - Conversion of vitamin D to a more active form
  - Regulation of calcium and phosphorus balance by filtering and reabsorbing approximately half of unbound serum calcium
  - Activation of vitamin D_3, a compound that promotes intestinal calcium absorption and regulates phosphate excretion

NEPHRON

- Basic structural and functional unit of kidneys
- Performs two main functions:
  - Mechanically filters fluids, wastes, electrolytes, acids, and bases into the tubular system
  - Selectively reabsorbs and secretes ions

- Each consists of a *glomerulus* (cluster of capillaries located inside Bowman's capsule) and a collecting duct
- Divided into three portions:
  - Proximal convoluted tubule
  - Loop of Henle (which, with its accompanying blood vessels and collecting tubules, forms the renal pyramids in the medulla)
  - Distal convoluted tubule

FILTRATE MOVEMENT THROUGH THE NEPHRON

- The proximal convoluted tubule has freely permeable cell membranes that allow reabsorption of nearly all glucose, amino acids, metabolites, and electrolytes from filtrate and allow the circulation of large amounts of water
- When the filtrate enters the descending limb of the loop of Henle, its water content has been reduced by 70% and it contains a high concentration of salts, chiefly sodium
- As the filtrate moves deeper into the medulla and the loop of Henle, osmosis draws even more water into the extracellular spaces, further concentrating the filtrate
- After the filtrate enters the ascending limb, its concentration is readjusted by the transport of ions into the tubule
- Transport continues until the filtrate enters the distal convoluted tubule

Ureters

- Fibromuscular tubes that carry urine from the kidneys to the bladder with the aid of peristaltic waves
- Left is usually slightly longer than the right, so the left kidney is usually higher than the right
- Surrounded by a three-layered (mucosa, muscularis, and fibrous coat) wall

Bladder

- Hollow, sphere-shaped, muscular organ in the pelvis
- Stores urine (capacity ranges from 500 to 600 ml)
- Base contains three openings that form a triangular area called the *trigone*
- Two of the openings connect the bladder to the ureters
- Third opening connects the bladder to the urethra
- When urine fills the bladder, parasympathetic nerve fibers in the bladder wall cause the bladder to contract and the internal sphincter to relax

In a voluntary reaction, the cerebrum causes the external sphincter to relax and urination to begin (called the *micturition reflex*)

Urethra

- Small duct that channels urine from the bladder out of the body
- In females:
  - Embedded in the anterior wall of the vagina behind the symphysis pubis
  - Connects the bladder with an external opening, or urethral meatus, located anterior to the vaginal opening
- In males:
  - Passes vertically through the prostate gland and then extends through the urogenital diaphragm and the penis
  - Serves as a passageway for semen as well as urine

Urine

- Formation results from three processes that occur in the nephrons: glomerular filtration, tubular reabsorption, and tubular secretion
- Consists of sodium, chloride, potassium, calcium, magnesium, sulfates, phosphates, bicarbonates, uric acid, ammonium ions, creatinine, and urobilinogen (a derivative of bilirubin resulting from the action of intestinal bacteria)
- Composition of excreted form is determined by substances absorbed and secreted in the nephrons
- Total daily output averages 720 to 2,400 ml (varies with fluid intake and climate)

Hormones and the urinary system

- Hormones help the body manage tubular reabsorption and secretion

Antidiuretic hormone

- Regulates urine output
  - High levels of antidiuretic hormone (ADH) increase water absorption and urine concentration
  - Low levels of ADH decrease water absorption and dilute the urine

Renin-angiotensin system

- Effects blood pressure
  - Renin leads to the formation of a hormone called *angiotensin I*
  - As angiotensin I circulates through the lungs, it's converted into angiotensin II by angiotensin-converting enzyme
  - Angiotensin II exerts a powerful constrictor effect on the arterioles, thus raising blood pressure
- Primary function is to maintain blood pressure in such situations as hemorrhage and extreme salt depletion

- Factors that stimulate the kidneys to release renin:
  - Low blood pressure
  - Low salt levels in filtrate passing through the kidneys

Aldosterone

- Secreted by the adrenal gland in response to stimulation by the renin-angiotensin system
- Facilitates tubular reabsorption by regulating sodium retention and helping to control potassium secretion by epithelial cells in the tubules
- Secretion by adrenal cortex increases in response to increased serum potassium levels, causing sodium retention and thereby raising blood pressure

Other hormones

- Erythropoietin: Secreted by kidneys in response to low arterial oxygen tension to stimulate increased RBC production

Cross-training

Test your knowledge of terms related to the urinary system by completing the following crossword puzzle.

Across

1. The urethra passes vertically through this gland in males
5. Small duct that channels urine from the bladder to the outside of the body
7. The kidney's functional unit
10. Fibromuscular tubes that connect each kidney to the bladder
11. A derivative of bilirubin resulting from the action of intestinal bacteria
12. Hollow, sphere-shaped, muscular organ in the pelvis

Down

2. Hormone secreted by the kidneys in response to low arterial oxygen tension
3. Reflex that causes the external sphincter to relax and urination to begin
4. Circular contraction and relaxation of a tube-shaped structure
6. Tubular apparatus inside the nephron
8. Name of the gland that lies on top of each kidney
9. Enzyme secreted by the kidneys

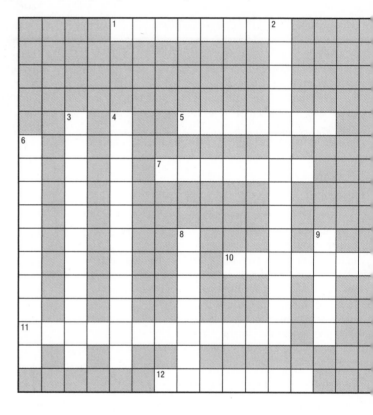

I like to play it safe. I'm protected in front by the contents of the abdomen and behind by the muscles attached to the vertebral column. Then, just to be sure, a surrounding layer of fat offers a bit more cushion.

Train your brain

Sound out the pictures and symbols to discover a fact about the urinary system.

The [boy] + [kidneys] R C-shaped,
h + [eye] + ly vas + Q + lar [organ] + s.

Strike out

The kidneys perform many functions. However, some of the functions listed below aren't performed by the kidneys. Cross out all the functions not performed by the kidneys.

1. Eliminate wastes and excess ions in the form of urine
2. Regulate the chemical composition of blood by producing sodium, potassium, and phosphorus
3. Influence blood volume
4. Help maintain fluid-electrolyte and acid-base balances
5. Help maintain glucose levels by stimulating the release of insulin from the pancreas
6. Secrete enzymes that govern blood pressure
7. Convert vitamin A to a more active form
8. Secrete a hormone that stimulates RBC production

Talk about covering all the bases! The kidneys perform functions ranging from eliminating wastes to stimulating RBC production.

■■
■ Finish line

Label the kidney structures shown in this illustration.

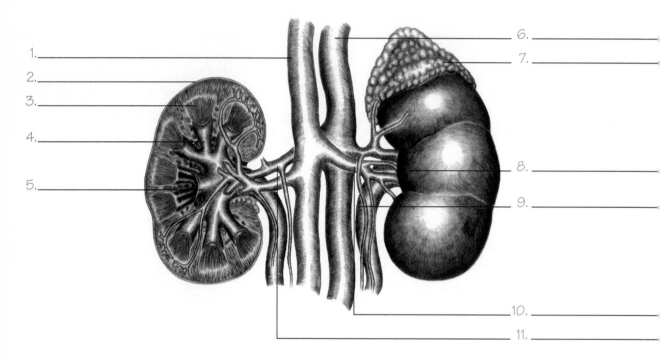

1. _____

2. _____

3. _____

4. _____

5. _____

6. _____

7. _____

8. _____

9. _____

10. _____

11. _____

■■
■ Hit or miss

Some of the following statements about the kidneys are true; the others are false. Mark each accordingly.

_____ 1. The left kidney is situated slightly lower than the right to make room for the pancreas.

_____ 2. The kidneys are fixed securely in place by muscles attached to the vertebral column.

_____ 3. The kidneys receive waste-filled blood from the renal artery, which branches off the abdominal aorta.

_____ 4. The renal artery subdivides into several branches when it enters the kidney.

_____ 5. Filtered blood returns to the circulation by way of the renal vein, which empties into the superior vena cava.

_____ 6. The nephrons remove waste products from the blood, which are then excreted in urine.

_____ 7. The kidneys receive about 20% of the blood pumped by the heart each minute.

Power stretch

Unscramble the words below to discover the names of the three regions of the kidneys. Then draw lines to link each region to its particular characteristics.

LEARN TREXOC

— — — — —

— — — — — —

NEARL LAUDELM

— — — — —

— — — — — — —

REALN LEVIPS

— — — — —

— — — — — —

A. Inner region of the kidney

B. Contains blood-filtering mechanisms

C. Functions as the kidney's collecting chamber

D. Outer region of the kidney

E. Protected by a fibrous capsule and layers of fat

F. Middle region of the kidney

G. Receives urine through the major calyces

Match point

Match each of the structures listed on the left to its function on the right.

1. Nephron _____
2. Renal artery _____
3. Interlobular artery _____
4. Afferent arteriole _____
5. Glomerulus _____
6. Efferent arteriole _____

A. Runs between the lobes of the kidneys

B. Conveys blood away from the glomerulus

C. Conveys blood to the glomerulus

D. Basic structural and functional unit of the kidney

E. Tubular apparatus inside the nephron

F. Carries blood to each kidney

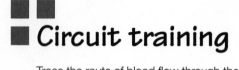

Trace the route of blood flow through the kidneys by drawing arrows between the appropriate boxes.

Renal artery

Abdominal aorta	Renal vein	Efferent arterioles

Inferior vena cava	Interlobular artery	Glomerulus

Peritubular capillaries	Afferent arterioles

Finish line

This illustration shows the structure of the nephron. See if you can name all of its components.

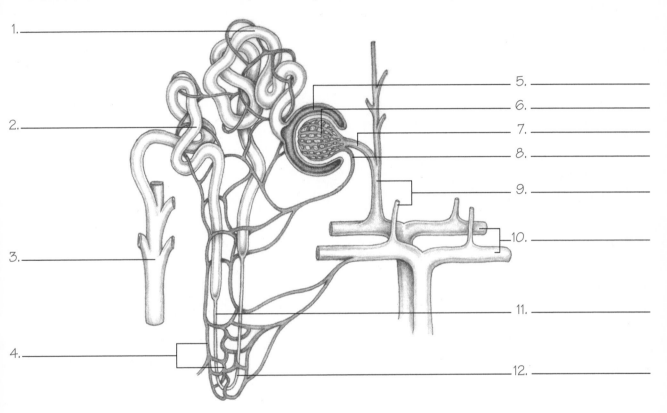

1. _____

2. _____

3. _____

4. _____

5. _____

6. _____

7. _____

8. _____

9. _____

10. _____

11. _____

12. _____

> Nephrons perform two main functions. First, they mechanically filtrate fluids, wastes, electrolytes, acids, and bases into the tubular system. Second, they selectively reabsorb and secrete ions, allowing precise control of fluid and electrolyte balance.

Jumble gym

Unscramble the following words to discover the portions of the nephron. Then use the circled letters to answer the question posed.

Question: **Where are the glomeruli located?**

1. L E G L O U R S U M

   _ _ _ _ _ ⊖ _ _ _

2. M O L A R P I X   C O O L D U N V E T   B L U T U E

   _ ⊖ _ ⊖ _ _ _   _ _ _ _ _ ⊖ _ _ _ _   _ _ _ _ _ _ ⊖

3. P O L O   F O   L E E H N

   _ ⊖ _ _   _ _   _ _ ⊖ _ _

4. S I T D A L   D U C T L O V E O N   L U B E T U

   _ _ _ _ ⊖ _   ⊖ _ _ _ _ _ _ ⊖ _ _   _ _ _ _ _ ⊖

Answer: __ __ __ __   __ __ __ __ __ __

Train your brain

Sound out the pictures and symbols to discover a fact about the urinary system.

The loops of 🐔 +le, along with their 🩸 🚢 +s and collecting 🧪 + ules, 4 +m the renal 🔺 +s.

■ Hit or miss

Some of the following statements regarding urine formation are true; the others are false. Mark each accordingly.

_____ 1. The proximal convoluted tubules have freely permeable cell membranes.

_____ 2. In the proximal convoluted tubules, nearly all of the filtrate's glucose, amino acids, metabolites, and electrolytes are reabsorbed into nearby capillaries.

_____ 3. When the filtrate enters the descending limb of the loop of Henle, it still contains a large amount of water.

_____ 4. At this point, nearly all of the filtrate's sodium has been reabsorbed.

_____ 5. As the filtrate moves deeper into the medulla and the loop of Henle, osmosis draws more water into the extracellular spaces, further concentrating the filtrate.

_____ 6. After the filtrate enters the ascending limb, its concentration is readjusted by the transport of ions into the tubule.

_____ 7. The transport of ions continues until the filtrate enters the distal convoluted tubule.

Coaching session
Urinary changes with aging

Kidneys
- Decreased function after age 40 (by as much as 50% by age 90)
- Decreased size
- Decreased glomerular filtration rate
- Decreased blood flow
- Decreased ability to respond to variations in sodium intake
- Reduced number of functioning nephrons, resulting in decreased tubular reabsorption and renal concentrating ability
- Impaired clearance of drugs

Bladder
- Muscle weakening
- Reduced size and capacity
- Incomplete emptying (increasing the risk of infection)

Other
- Increased residual urine, frequency of urination, and nocturia

■ Match point

Match each of the terms below left with its definition.

1. Ureters _____
2. Bladder _____
3. Trigone _____
4. Urethra _____
5. Urobilinogen _____

A. A derivative of bilirubin normally found in urine
B. Triangular area at the base of the bladder containing three openings
C. Fibromuscular tubes that carry urine from the kidneys to the bladder
D. Hollow, sphere-shaped, muscular organ in the pelvis that collects urine
E. Small duct that channels urine from the bladder to outside the body

■ Batter's box

Fill in the blanks with the appropriate words.

Urine or you're out

In a normal adult, bladder capacity ranges from _____ to _____
                                                        1   2
ml. If the amount of urine exceeds bladder capacity, the bladder distends above the

_____ . The base of the bladder contains three openings: two of the openings
         3

connect the bladder to the _____ ; the third connects the bladder to the
                                    4

_____ . Urination results from _____ (reflex) and
         5   6

_____ (learned or intentional) processes. When urine fills the bladder,
         7

_____ nerve fibers in the bladder wall cause the bladder to contract and the
         8

_____ sphincter (located at the internal urethral orifice) to relax. The
         9

_____ , in a voluntary reaction, then causes the _____ sphinc-
        10   11

ter to relax and urination to begin. The reflex that causes urination is called the

_____ reflex.
        12

Options

500

600

cerebrum

external

internal

involuntary

micturition

parasympathetic

symphysis pubis

ureters

urethra

voluntary

Jumble gym

Unscramble the following words to discover the names of hormones that affect the urinary system.
Then use the circled letters to answer the question posed.

Question: **What's the name of the enzyme secreted by the kidneys that leads to the formation of angiotensin I?**

1. C U R T A I N E D I T I   M E N O H O R

   — — — — — — — — — —   — — — — — ⊖ —

2. E A T I N G I N S O N   I

   — ⊖ — — — — — — —   —

3. S A N G N I T E I N O   I I

   — — — — — — — — ⊖ —   — —

4. T R A D E N O E L S O

   — — — — — — ⊖ — —

5. H E R O I N E T R I P T O Y

   — — — — — — — — — ⊖ — — —

Answer: __ __ __ __ __

Hit or miss

Some of the following statements are true; the others are false. Mark each accordingly.

_____ 1. High levels of ADH decrease water absorption and dilute urine.

_____ 2. Renin leads to the formation of the hormone angiotensin I.

_____ 3. Angiotensin I exerts a powerful constricting effect on the arterioles, leading to an elevation in blood pressure.

_____ 4. The primary function of the renin-angiotensin system is to maintain blood pressure in situations such as hemorrhage and extreme sodium depletion.

_____ 5. The renin-angiotensin system also acts on the adrenal gland to release aldosterone.

_____ 6. Aldosterone regulates potassium retention, which helps maintain blood pressure.

_____ 7. The hormone erythropoietin stimulates RBC production in response to low arterial oxygen tension.

You make the call

Using the spaces provided describe the three steps of urine formation.

Step 1: Glomerular filtration

_____

_____

_____

_____

_____

Step 2: Tubular reabsorption

_____

_____

_____

_____

_____

Step 3: Tubular secretion

_____

_____

_____

_____

■ Mind sprints

Go the distance by listing as many of the substances normally found in urine as you can in 2 minutes.

1. _____
2. _____
3. _____
4. _____
5. _____
6. _____
7. _____
8. _____
9. _____
10. _____
11. _____
12. _____
13. _____
14. _____
15. _____

■ You make the call

ADH regulates fluid balance in four steps. Using the illustration below as a guide, describe each of these steps.

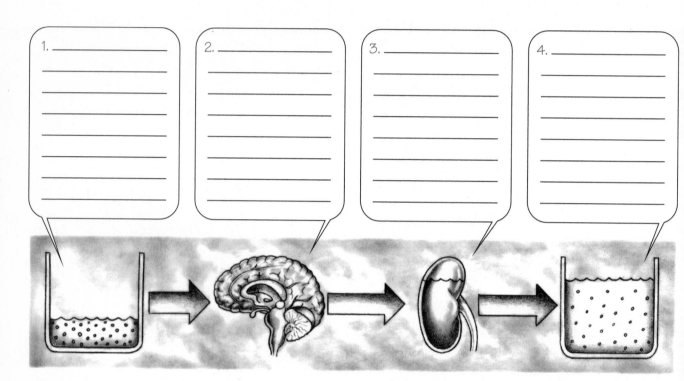

1. _____

2. _____

3. _____

4. _____

■ You make the call

The renin-angiotensin system triggers the release of aldosterone from the adrenal gland. Using the illustration below as a guide, describe the steps leading up to aldosterone release.

1. _____
_____
_____
_____
_____

4. _____
_____
_____
_____
_____

6. _____
_____
_____
_____
_____

8. _____
_____
_____
_____
_____

2. _____
_____
_____
_____
_____

5. _____
_____
_____
_____

7. _____
_____
_____
_____

3. _____
_____
_____
_____

■ Starting lineup

Test your knowledge of how the renin-angiotensin system regulates blood pressure by putting the following steps in the correct order.

Angiotensin II constricts the arterioles.

Renin causes the formation of angiotensin I.

Renin is secreted by the kidneys and is circulated in blood.

Blood pressure rises.

Angiotensin I is converted into angiotensin II.

1.

2.

3.

4.

5.

The renin-angiotensin system packs a one-two punch when it comes to raising blood pressure.

Fluids, electrolytes, acids, and bases

Proper balance of fluids, electrolytes, acids, and bases is necessary for a healthy body. See if your knowledge of this topic is properly balanced by completing this next obstacle.

Fluids, electrolytes, acids, and bases review

Overview

- Health and homeostasis (equilibrium of the various body functions) of the human body depend on fluid, electrolyte, and acid-base balance
- Factors that disrupt this balance, such as surgery, illness, and injury, can lead to potentially fatal changes in metabolic activity

Fluid balance

- Fluids are made up of water that contains solutes (dissolved substances) that are necessary for physiologic functioning
  - Examples of solutes include electrolytes, glucose, amino acids, and other nutrients
- Intracellular fluid (ICF): Found within the individual cells of the body; represents about 40% of adult's total body weight
- Intravascular fluid (IVF): Found within the plasma and lymphatic system
- Interstitial fluid (ISF): Found in the loose tissue around cells
- Extracellular fluid (ECF): Found in the spaces between cells; a combination of IVF and ISF; represents about 20% of an adult's total body weight

Types of solutions

ISOTONIC SOLUTION
- Has the same solute concentration as another solution
- No imbalance means no net fluid shift

HYPOTONIC SOLUTION
- Has a lower solute concentration than another solution
- Fluid from the hypotonic solution shifts into the second solution until the two solutions have equal concentrations

HYPERTONIC SOLUTION
- Has a higher solute concentration than another solution
- Fluid is drawn from the second solution into the hypertonic solution until the two solutions have equal concentrations

Fluid movement within cells

- Fluids and solutes move constantly within the body to maintain homeostasis
- Solutes within the various compartments of the body move through semipermeable (allow some solutes to pass through but not others) membranes that separate those compartments

DIFFUSION
- Substances move from an area of higher concentration to an area of lower concentration
- Movement continues until distribution is uniform

ACTIVE TRANSPORT
- Solutes move from an area of lower concentration to an area of higher concentration
- Adenosine triphosphate (ATP) supplies energy for solute movement in and out of cells
- Sodium and potassium use ATP to move in and out of cells in a form of active transport called the *sodium-potassium pump*

OSMOSIS
- Passive movement of fluid across a membrane from an area of lower solute concentration and relatively more fluid into an area of higher solute concentration and relatively less fluid
- Stops when enough fluid has moved through the membrane to equalize the solute concentration on both sides of the membrane

WATER BALANCE
- Water normally enters the body from the GI tract
- The body obtains about 1.5 L (1.6 qt) of water daily from consumed liquids and obtains another approximately 800 ml (26.6 oz) from solid foods
- Oxidation of food in the body yields carbon dioxide (CO_2) and about 300 ml (10 oz) of water (water of oxidation)
- Water leaves the body through the skin (in perspiration), lungs (in expired air), GI tract (in feces), and urinary tract (in urine)
- The main route of water loss is urine excretion, which typically varies from 1 to 2.6 L daily
- Water losses through the skin and lungs amount to 1 L daily but may increase markedly with strenuous exertion
- Thirst is the primary regulator of fluid intake
  - When the body becomes dehydrated, ECF volume is reduced, causing an increase in sodium concentration and osmolarity
  - When sodium concentration reaches about 2 mEq/L above normal, neurons of the thirst center in the hypothalamus are stimulated
  - The brain then directs motor neurons to satisfy thirst, causing the person to drink enough fluid to restore ECF to normal
  - Through the countercurrent mechanism, the kidneys regulate fluid output by excreting urine of greater or lesser concentration, depending on fluid balance

Electrolyte balance

▧ Electrolytes: substances that break up into electrically charged particles (ions) when dissolved in water.
▧ Positively charged cations
– Sodium
– Potassium
– Calcium
– Magnesium
▧ Negatively charged anions
– Chloride
– Bicarbonate (HCO_3^-)
– Phosphate
▧ Normally, the electrical charges of cations balance the electrical charges of anions, keeping body fluids electrically neutral
▧ Electrolytes profoundly affect the body's water distribution, osmolarity, and acid-base balance
▧ Numerous mechanisms within the body help maintain electrolyte balance
▧ Dysfunction or interruption of any of these mechanisms can produce an electrolyte imbalance

Regulatory mechanisms for common electrolytes

SODIUM REGULATORS

▧ Kidneys and aldosterone are chief regulators
▧ Small intestine absorbs sodium readily from food
▧ Skin and kidneys excrete sodium

POTASSIUM REGULATORS

▧ Kidneys control potassium through aldosterone action
▧ Most potassium is absorbed from food in the GI tract
▧ Normally, the amount excreted in urine equals dietary potassium intake

CALCIUM REGULATORS

▧ Parathyroid hormone (PTH)
– Main regulator
– Controls both calcium uptake from the GI tract and calcium excretion by the kidneys

MAGNESIUM REGULATORS

▧ Aldosterone controls renal magnesium reabsorption
▧ Magnesium is absorbed from the GI tract and excreted in urine, breast milk, and saliva

CHLORIDE REGULATORS

▧ Kidneys
▧ Chloride ions move in conjunction with sodium ions

BICARBONATE REGULATORS

▧ Kidneys regulate bicarbonate by excreting, absorbing, or forming it
▧ Bicarbonate, in turn, plays a vital part in acid-base balance

PHOSPHATE REGULATORS

▧ Kidneys
▧ Absorbed from food
▧ PTH governs phosphate levels

Acid-base balance

▧ Stable concentration of hydrogen ions in body fluids
▧ Hydrogen ion concentration of a fluid determines whether it's acidic or basic (alkaline)
▧ Acid—a substance that yields hydrogen ions when changed from a complex to a simpler compound in solution
▧ Base—dissociates in water, releasing ions that can combine with hydrogen ions
▧ The body produces acids, thus yielding hydrogen ions, through the following mechanisms:
– Protein catabolism yields nonvolatile acids, such as sulfuric, phosphoric, and uric acids
– Fat oxidation produces acid ketone bodies
– Anaerobic glucose catabolism produces lactic acid
– Intracellular metabolism yields carbon dioxide as a by-product; carbon dioxide dissolves in body fluids to form carbonic acid (H_2CO_3)
▧ Blood pH stays within a narrow range: 7.35 to 7.45 and is maintained by buffer systems and the lungs and kidneys, which neutralize and eliminate acids as rapidly as they're formed

Buffer systems

▧ Reduce the effect of an abrupt change in hydrogen ion concentration by converting a strong acid or base into a weak acid or base (which releases fewer hydrogen ions)
▧ Buffer systems that help maintain acid-base balance include:
– Sodium bicarbonate–carbonic acid
– Phosphate
– Protein

SODIUM BICARBONATE–CARBONIC ACID BUFFER SYSTEM

▧ Major buffer in ECF
▧ Sodium bicarbonate ($NaHCO_3^-$) concentration is regulated by the kidneys
▧ Carbonic acid concentration is regulated by the lungs
▧ The strong base (sodium hydroxide [$NaOH$]) is replaced by sodium bicarbonate and water (H_2O)
▧ Sodium hydroxide dissociates almost completely and releases large amounts of hydroxyl (OH^-)
▧ If a strong acid is added, the opposite occurs

PHOSPHATE BUFFER SYSTEM

▧ Regulates pH of fluids as they pass through the kidneys
▧ Acidic component is sodium dihydrogen phosphate (NaH_2PO_4)
▧ Its alkaline component is sodium monohydrogen phosphate (Na_2HPO_4)

PROTEIN BUFFER SYSTEM

■ Intracellular proteins absorb hydrogen ions generated by the body's metabolic processes

Lungs

■ Respiration plays a crucial role in controlling pH
– The protein buffer system changes the pH of the blood in 3 minutes or less by changing the breathing rate
– Decreased respiratory rate decreases the exchange and release of carbon dioxide, resulting in decreased hydrogen (H+) and increased pH
– The lungs excrete carbon dioxide and regulate the carbonic acid content of the blood
– A change in the rate or depth of respirations can alter the carbon dioxide content of alveolar air and the alveolar partial pressure of carbon dioxide (PCO_2)
■ An increase in alveolar PCO_2 raises the blood concentration of carbon dioxide and carbonic acid, which in turn stimulates the respiratory center to increase respiratory rate and depth
– As a result, alveolar PCO_2 decreases, which leads to a corresponding drop in carbonic acid and carbon dioxide concentrations in the blood
■ A decrease in respiratory rate and depth has the reverse effect (raises alveolar PCO_2, which in turn triggers an increase in the blood's carbon dioxide and carbonic acid concentrations)

Kidneys

■ Help manage acid-base balance by regulating the blood's bicarbonate concentration

RENAL TUBULAR ION SECRETION

■ Each hydrogen ion secreted into the tubular filtrate joins with a bicarbonate ion to form carbonic acid, which rapidly dissociates into carbon dioxide and water
■ Carbon dioxide diffuses into the tubular epithelial cell, where it combines with more water and forms additional carbonic acid.

BICARBONATE REABSORPTION

■ Remaining water molecules in the tubular filtrate are eliminated in urine
■ As each hydrogen ion enters the tubular filtrate to combine with a bicarbonate ion, a bicarbonate ion in the tubular epithelial cell diffuses into the circulation

FORMATION OF AMMONIA AND PHOSPHATE SALTS

■ To form more bicarbonate, the kidneys must secrete additional hydrogen ions in exchange for sodium ions
■ For the renal tubules to continue secreting hydrogen ions, the excess ions must combine with other substances in the filtrate and be excreted
■ Excess hydrogen ions in the filtrate may combine with ammonia produced by the renal tubules or with phosphate salts present in the tubular filtrate
■ Ammonia forms in the tubular epithelial cells by removal of the amino groups from glutamine and from amino acids delivered to the tubular epithelial cells from the circulation

■ After diffusing into the filtrate, ammonia joins with the secreted hydrogen ions, forming ammonium ions
■ These ions are excreted in the urine with chloride and other anions; each secreted ammonia molecule eliminates one hydrogen ion in the filtrate
■ At the same time, sodium ions that have been absorbed from the filtrate and exchanged for hydrogen ions enter the circulation, as does the bicarbonate formed in the tubular epithelial cells
■ Some secreted hydrogen ions combine with sodium monohydrogen phosphate
■ Each of the secreted hydrogen ions that joins with the disodium salt changes it to the monosodium salt sodium dihydrogen phosphate
■ The sodium ion released in this reaction is absorbed into the circulation along with a newly formed bicarbonate ion

FACTORS AFFECTING BICARBONATE FORMATION

■ The rate of bicarbonate formation by renal tubular epithelial cells is affected by two factors:
– Amount of dissolved carbon dioxide in the plasma
– Potassium content of the tubular cells
■ If the plasma carbon dioxide level rises, renal tubular cells form more bicarbonate
■ Increased plasma carbon dioxide encourages greater carbonic acid formation by renal tubular cells

Batter's box

Fill in the blanks with the appropriate words. *Hint:* Answers may be used more than once.

Balancing act

The health of the human body depends on fluid, electrolyte, and acid-base

_____ . Factors that disrupt this _____ , such as surgery,

1   2

illness, and injury, can lead to potentially fatal changes in _____

3

activity.

      The constant movement of fluid and electrolytes allows the body to maintain

_____ , the state of balance the body seeks. Each day, the body takes in

4

_____ from the GI tract. It also loses fluid through the

5

_____ , lungs, intestines, and _____ tract. The normal

6   7

daily fluid intake is _____ to _____ in liquids,

8   9

_____ in solid foods, and _____ in water of oxidation.

10   11

Total daily fluid gains should _____ daily fluid losses.

12

Salute to solutes

Solutes move between the various compartments of the body by passing through

_____ membranes. In _____ , solutes move from an area

13   14

of high concentration to an area of lower concentration. In _____ ,

15

solutes move from an area of lower concentration to an area of higher concentration.

The energy required for a solute to move against a concentration gradient comes from

a substance called _____ .

16

Options

300 ml

800 ml

1.5 L

2.6 L

active transport

adenosine triphosphate

balance

diffusion

equal

fluid

homeostasis

metabolic

semipermeable

skin

urinary

Cross-training

Test your knowledge of terms related to fluids, electrolytes, acids, and bases by completing the crossword puzzle below.

Across

2. Solution with a pH greater than 7
5. Negatively charged ions
9. Positively charged ions
11. A substance that yields hydrogen ions when dissociated in solution
12. Constant state of balance
13. Buffer system that can change the pH of the blood in 3 minutes or less by changing the breathing rate

Down

1. Primary regulator of fluid intake
3. Electrically charged particles
4. Substances that dissociate into electrically charged particles when dissolved in water
6. Substances dissolved in water
7. Buffer system that works by regulating the pH of fluids as they pass through the kidneys
8. The passive movement of fluid across a membrane from an area of lower solute concentration, and comparatively more fluid, into an area of higher solute concentration, and comparatively less fluid
10. Concentration of this ion determines whether a solution is acidic or basic

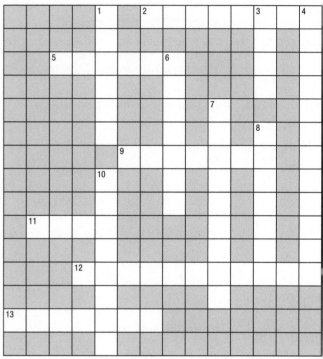

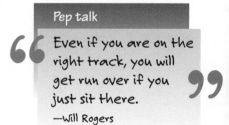

Pep talk

Even if you are on the right track, you will get run over if you just sit there.

—Will Rogers

■ Match point

Match each type of body fluid listed below with its correct definition.

1. ICF _____
2. IVF _____
3. ISF _____
4. ECF _____

A. Fluid found in the spaces between cells

B. Fluid found within the plasma and the lymphatic system

C. Fluid found within the individual cells of the body

D. Fluid found in the loose tissue around cells

Fluid is constantly on the move within the body.

■ Power stretch

Unscramble the words below to discover three different types of solutions. Then draw lines linking each fluid type (which, for the sake of this puzzle, is known as the "first solution") to its particular characteristics and the type of fluid shift it causes.

COINSITO

_ _ _ _ _ _ _ _ _

TYCOONHIP

_ _ _ _ _ _ _ _ _

PITCHERYON

_ _ _ _ _ _ _ _ _ _

A. Causes fluid to shift from the first solution into the second solution until the two solutions have equal concentrations

B. Causes fluid to be drawn from the second solution into the first solution until the two solutions have equal concentration

C. No fluid shift takes place

D. Has a lower solute concentration than another solution

E. Has the same concentration as another solution

F. Has a higher solute concentration than another solution

Knowing the different types of solutions is all about concentration.

■ Interval training

Test your knowledge of the body's major electrolytes by filling in the blanks in the chart below.

Electrolyte	ICF	ECF
Cations		
_____1	10 mEq/L	136 to 146 mEq/L
Potassium	140 mEq/L	_____2
_____3	10 mEq/L	4.5 to 5.8 mEq/L
Magnesium	_____4	1.6 to 2.2 mEq/L
Anions		
Chloride	_____5	96 to 106 mEq/L
_____6	10 mEq/L	24 to 28 mEq/L
_____7	100 mEq/L	_____8

> Normally, the electrical charges of cations balance the electrical charges of anions, keeping body fluids electrically neutral.

Starting lineup

Test your knowledge of the body's osmotic regulation of decreased sodium level or water excess by putting the following steps in the correct order.

Serum osmolality drops to less than 280 mOsm/kg.	1.
Renal water excretion increases.	2.
Antidiuretic hormone release is suppressed.	3.
Serum osmolality normalizes.	4.
Thirst decreases.	5.

Power stretch

Unscramble the words on the left to discover the names of buffer systems that help maintain acid-base balance. Then draw lines linking each system to its particular characteristics.

A. The major buffer in ECF

B. Can exist in the form of acids or alkaline salts

C. Regulates the pH of fluids as they pass through the kidneys

D. Can alter breathing rate to change blood pH

E. Uses the kidneys to regulate sodium bicarbonate concentration and the lungs to regulate carbonic acid

F. Has sodium dihydrogen phosphate as its acidic component and sodium monohydrogen phosphate as its alkaline component

G. Replaces the strong base (sodium hydroxide) with sodium bicarbonate and water; sodium hydroxide dissociates and releases large amounts of hydroxyl

H. Uses intracellular proteins to absorb hydrogen ions

DOSIUM
ACABBIETORN–RANBOCCI
CADI

_ _ _ _ _ _
_ _ _ _ _ _ _ _ _ _ _ _ _ –
_ _ _ _ _ _ _ _ _
_ _ _ _

HEATHPOPS

_ _ _ _ _ _ _ _ _

POINTER

_ _ _ _ _ _ _

Yes, I am buff. But enough about me. Buffer systems reduce the effect of an abrupt change in hydrogen ion concentration by converting a strong acid or base into a weak acid or base.

■ ■
■ Hit or miss

Some of the following statements are true; the others are false. Mark each accordingly.

_____ 1. A rise in the carbon dioxide content of arterial blood or a decrease in blood pH suppresses the respiratory system.

_____ 2. When the respiratory rate increases, less carbon dioxide, and therefore, less carbonic acid and fewer hydrogen ions remain in the blood; consequently, blood pH increases.

_____ 3. The kidneys help manage acid-base balance by regulating the blood's bicarbonate concentration.

_____ 4. Recovery and formation of bicarbonate in the kidneys depend on sodium excretion by the renal tubules.

_____ 5. One of the factors influencing the rate of bicarbonate formation by renal tubular epithelial cells is the amount of dissolved carbon dioxide in the plasma.

_____ 6. The other factor influencing the rate of bicarbonate formation by renal tubular epithelial cells is the sodium content of the tubular cells.

As hydrogen ion concentration increases, pH decreases.

Coaching session
Understanding pH

Hydrogen ion concentration is commonly expressed as pH, which indicates the degree of acidity or alkalinity of a solution.

pH Facts
- pH of 7 = neutral (equal amounts of hydrogen and hydroxyl ions)
- pH < 7 = acidic (more hydrogen ions than hydroxyl ions)
- pH > 7 = alkaline (less hydrogen ions than hydroxyl ions)

Team colors

In the following diagram, color the squares containing information about the factors leading to respiratory alkalosis red. Color the squares containing information about the factors leading to respiratory acidosis blue.

CO_2 is retained

Pulmonary ventilation increases	Excessive CO_2 is exhaled	pH rises above 7.45

Pulmonary ventilation decreases	Pa_{CO_2} declines	pH falls below 7.35

Retained CO_2 combines with water to form excessive H_2CO_3	Pa_{CO_2} level rises above 45 mm Hg

H_2CO_3 dissociates to release free H and HCO_3^- ions	Pa_{CO_2} level falls below 35 mm Hg

H_2CO_3 production declines

H and HCO_3^- ions are lost

■ Starting lineup

Test your knowledge of what happens in metabolic alkalosis and acidosis by putting the following physiologic processes in the correct order.

Metabolic alkalosis

Kidneys excrete excess sodium ions, water, and bicarbonate to maintain electrochemical balance.	1.
Hydrogen ions diffuse out of cells, and potassium ions move into cells.	2.
$Paco_2$ increases.	3.
Bicarbonate that can't be reabsorbed by the renal glomeruli is excreted in urine.	4.
Excess bicarbonate elevates serum pH levels above 7.45, leading to decreased respiratory rate.	5.
Calcium ionization decreases, allowing sodium ions to overstimulate nerve cells.	6.
Chemical buffers bind with accumulated bicarbonate ions.	7.

Metabolic acidosis

$Paco_2$ decreases.	1.
Excess hydrogen decreases pH levels below 7.35, leading to increased respiratory rate.	2.
Chemical buffers bind with accumulated hydrogen ions.	3.
Excess hydrogen ions lead to reduced excitability of nerve cells, causing central nervous system depression.	4.
Kidneys secrete excess hydrogen ions into the renal tubules, which are then excreted in urine.	5.
Hydrogen ions diffuse into cells, and potassium ions move into the blood.	6.

15

Reproductive system

■ Warm-up

Reproductive system review

Male reproductive system

- Consists of the organs that produce, transfer, and introduce mature sperm into the female reproductive tract, where fertilization occurs
- Supplies male sex cells (spermatogenesis)
- Plays a part in the secretion of male sex hormones

Penis

- Organ of copulation
- Deposits sperm in the female reproductive tract
- Acts as the terminal duct for the urinary tract
- Conduit for urine elimination
- Consists of an attached root, a free shaft, and an enlarged tip
- Major part formed by two corpora cavernosa
- On the underside, the corpus spongiosum encases the urethra
- Bulb formed by enlarged proximal end
- Glans penis:
  - Cone-shaped structure formed from the corpus spongiosum at the distal end of the shaft
  - Highly sensitive to sexual stimulation
- Urethral meatus opens through the glans to allow urination and ejaculation

Scrotum

- Extra-abdominal pouch that consists of a thin layer of skin overlying a tighter, musclelike layer
- Internally, a septum divides the scrotum into two sacs, which each contain a testis, an epididymis, and a spermatic cord

SPERMATIC CORD

- Connective tissue sheath that encases autonomic nerve fibers, blood vessels, lymph vessels, and the vas deferens
- Also called the *ductus deferens*
- Travels from the testis through the inguinal canal, exiting the scrotum through the external inguinal ring and entering the abdominal cavity through the internal inguinal ring

Testes

- Consists of two layers
  - Tunica vaginalis (outer layer)
  - Tunica albuginea (inner layer)

- Lobule contains one to four seminiferous tubules (small tubes in which spermatogenesis takes place)
- because sperm development requires a temperature lower than that of the rest of the body, the cremaster muscle elevates and lowers the testes to govern temperature

Duct system

- Consists of the epididymis, vas deferens, and urethra
- Conveys sperm from the testes to the ejaculatory ducts near the bladder

EPIDIDYMIS

- Coiled tube located superior to and along the posterior border of the testis
- During ejaculation, spermatozoa are ejected into the vas deferens

VAS DEFERENS

- Leads from the testes to the abdominal cavity, extends upward through the inguinal canal, arches over the urethra, and descends behind the bladder
- Ampulla (enlarged portion) merges with the duct of the seminal vesicle to form the short ejaculatory duct
- After passing through the prostate gland, joins with the urethra

URETHRA

- Small tube leading from the floor of the bladder to the exterior
- Consists of three parts
  - Prostatic urethra: surrounded by the prostate gland; drains the bladder
  - Membranous urethra: passes through the urogenital diaphragm
- Spongy urethra: makes up about 75% of the entire urethra

Accessory reproductive glands

- Produce the majority of semen

SEMINAL VESICLES

- Paired sacs at the base of the bladder

BULBOURETHRAL GLANDS

- Also known as *Cowper's glands*
- Paired and located inferior to the prostate

PROSTATE GLAND

- Walnut-sized
- Lies under the bladder and surrounds the urethra
- Consists of three lobes: the left and right lateral lobes and the median lobe
- Continuously secretes prostatic fluid
  - Thin, milky, alkaline fluid
  - During sexual activity, adds volume to semen
  - Enhances sperm motility
  - May aid conception by neutralizing the acidity of the urethra and the vagina

Semen

- Viscous, white secretion with a slightly alkaline pH (7.8 to 8)
- Consists of spermatozoa and accessory gland secretions
- Seminal vesicles produce roughly 60% of the fluid portion of the semen
- Prostate gland produces about 30% of the fluid portion of the semen

Spermatogenesis

- Sperm formation
- Begins when a male reaches puberty and normally continues throughout life

FIRST STAGE

- Primary germinal epithelial cells, called *spermatogonia*, grow and develop into primary spermatocytes
- Both spermatogonia and primary spermatocytes contain 46 chromosomes, consisting of 44 autosomes and the two sex chromosomes, X and Y

SECOND STAGE

- Primary spermatocytes divide to form secondary spermatocytes
- No new chromosomes are formed in this stage; the pairs only divide
- Each secondary spermatocyte contains one-half the number of autosomes, 22
- One secondary spermatocyte contains an X chromosome, the other, a Y chromosome

THIRD STAGE

- Each secondary spermatocyte divides again to form spermatids

FOURTH STAGE

- Spermatids undergo a series of structural changes that transform them into mature spermatozoa, or sperm
- Each spermatozoa has a head, neck, body, and tail
- Newly mature sperm pass from the seminiferous tubules through the vasa recta into the epididymis
- Most sperm move into the vas deferens, where they're stored until sexual stimulation triggers emission
- After ejaculation, sperm survive for 24 to 72 hours at body temperature

Hormonal control and sexual development

- Androgens (male sex hormones) are produced in the testes and the adrenal glands
- Androgens are responsible for the development of male sex organs and secondary sex characteristics
- Major androgens include testosterone, luteinizing hormone (LH), and follicle-stimulating hormone (FSH)

TESTOSTERONE

- Most significant male sex hormone
- Secreted by Leydig's cells
- Responsible for the development and maintenance of male sex organs and secondary sex characteristics
- Also required for spermatogenesis
- Secretion begins approximately 2 months after conception and directly affects sexual differentiation in the fetus
  - With testosterone, fetal genitalia develop into a penis, scrotum, and testes
  - Without testosterone, genitalia develop into a clitoris, vagina, and other female organs
- During the last 2 months of gestation, testosterone normally causes the testes to descend into the scrotum
- LH and FSH directly affect secretion of testosterone
- During puberty, the penis and testes enlarge and the male reaches full adult sexual and reproductive capability
- Puberty also marks the development of male secondary sexual characteristics
- After a male achieves full physical maturity, usually by age 20, sexual and reproductive function remain fairly consistent throughout life
- With aging, a man may experience subtle changes in sexual function but doesn't lose the ability to reproduce

Female reproductive system

- Unlike the male reproductive system, largely internal and housed within the pelvic cavity

External genitalia

MONS PUBIS

- Rounded cushion of fatty and connective tissue covered by skin and coarse, curly hair in a triangular pattern over the symphysis pubis

LABIA MAJORA

- Two raised folds of adipose and connective tissue that border the vulva on either side
- Extends from the mons pubis to the perineum

LABIA MINORA

- Two moist folds of mucosal tissue, dark pink to red in color, that lie within and alongside the labia majora
- Each upper section divides into an upper and lower lamella
  - The two upper lamellae join to form the prepuce (hood-like covering over the clitoris)
  - The two lower lamellae form the frenulum (posterior portion of the clitoris)
- Lower labial sections taper down and back from the clitoris to the perineum and join to form the fourchette (thin tissue fold along the anterior edge of the perineum)
- Contain sebaceous glands, which secrete a lubricant that also acts as a bactericide
- Swell in response to sexual stimulation, a reaction that triggers other changes that prepare the genitalia for coitus

CLITORIS

- Small, protuberant organ just beneath the arch of the mons pubis
- Contains erectile tissue, venous cavernous spaces, and specialized sensory corpuscles, which are stimulated during sexual activity

VESTIBULE

- Oval area bounded anteriorly by the clitoris, laterally by the labia minora, and posteriorly by the fourchette
- Skene's glands are found on both sides of the urethral opening
- Bartholin's glands are located laterally and posteriorly on either side of the inner vaginal orifice

URETHRAL MEATUS

- Slitlike opening below the clitoris through which urine leaves the body
- Center of the vestibule is the vaginal orifice

PERINEUM

- Located between the lower vagina and the anal canal
- Complex structure of muscles, blood vessels, fascia, nerves, and lymphatics

Internal genitalia

- Specialized organs whose main function is reproduction

VAGINA

- Highly elastic muscular tube located between the urethra and the rectum
- Wall has three tissue layers (epithelial tissue, loose connective tissue, and muscle tissue)
- Uterine cervix connects the uterus to the vaginal vault
- Three main functions:
  - Accommodates the penis during coitus
  - Channels blood discharged from the uterus during menstruation
  - Serves as the birth canal during childbirth

CERVIX

- Projects into the upper portion of the vagina
- Lower cervical opening is the *external os*
- Upper opening is the *internal os*
- Childbirth permanently alters the cervix

UTERUS

- Small, firm, pear-shaped, muscular organ situated between the bladder and rectum
- Typically lies at almost a 90-degree angle to the vagina
- Has a mucous membrane lining called the *endometrium*
- Has a muscular layer called the *myometrium*
- During pregnancy, the elastic, upper portion of the uterus, called the *fundus*, accommodates most of the growing fetus until term.
- The uterine neck joins the fundus to the cervix, the uterine part extending into the vagina
- The fundus and neck make up the corpus, the main uterine body

FALLOPIAN TUBES

- Attach to the uterus at the upper angles of the fundus
- Narrow cylinders of muscle fibers that are the site of fertilization
- Curved portion, called the *ampulla*, ends in the funnel-shaped infundibulum
- Fingerlike projections in the infundibulum, called *fimbriae*, move in waves that sweep mature ova from the ovary into the fallopian tube

OVARIES

- Located on either side of the uterus
- Size, shape, and position vary with age
- Take on an almond shape and a rough, pitted surface during the childbearing years
- After menopause, shrink and turn white
- Main function is to produce mature ova
- At birth, each ovary contains approximately 500,000 graafian follicles

▪ During the childbearing years, one graafian follicle produces a mature ovum during the first half of each menstrual cycle

▪ As the ovum matures, the follicle ruptures and the ovum is swept into the fallopian tube

▪ Also produce estrogen and progesterone as well as a small amount of androgens

Mammary glands

▪ Specialized accessory glands located in the breast that secrete milk

▪ Present in both sexes but typically function only in females

▪ Within the lobes are clustered acini (tiny, saclike duct terminals that secrete milk during lactation)

Hormonal function and the menstrual cycle

▪ Female body changes with age in response to hormonal control

▪ The hypothalamus, ovaries, and pituitary gland secrete hormones (estrogen, progesterone, FSH, and LH) that affect the buildup and shedding of the endometrium during the menstrual cycle

▪ During adolescence, the release of hormones causes a rapid increase in physical growth and spurs the development of secondary sex characteristics

▪ The menstrual cycle, which averages 28 days in length, is a complex process that involves the reproductive and endocrine systems

MENOPAUSE

▪ Cessation of menses (a woman is considered to have reached menopause after menses are absent for 1 year)

▪ Typically occurs between ages 40 and 55

▪ Includes the preceding 1 to 2 years of declining ovarian function, shown by irregular menses that gradually become further apart and produce a lighter flow

▪ Ovaries stop producing progesterone and estrogen

Cross-training

Test your knowledge of terms related to the reproductive system by completing the following crossword puzzle.

Across

4. Mucous membrane lining of the uterus
5. Curved portion of the fallopian tube
7. Highly elastic muscular tube located between the urethra and the rectum in females
9. Coiled tube located superior to and along the posterior border of the testis
10. Most significant male sex hormone
11. Viscous, white secretion containing spermatozoa

Down

1. Sperm formation
2. Muscular layer of the uterus
3. Male sex hormones
6. Male organ of copulation
8. Scrotal sac

◼◼
◼ Finish line

Identify the structures of the male reproductive system in the illustration below.

1. _____

2. _____

3. _____

4. _____

5. _____

6. _____

7. _____

8. _____

9. _____

10. _____

11. _____

12. _____

13. _____

14. _____

15. _____

16. _____

17. _____

18. _____

19. _____

20. _____

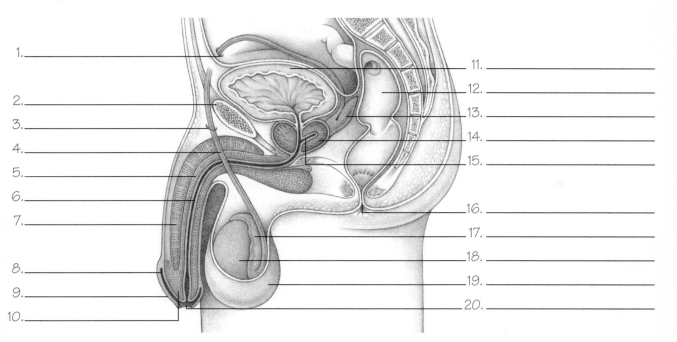

The three primary male sex hormones, or androgens, are testosterone, luteinizing hormone, and follicle-stimulating hormone.

■ Batter's box

Fill in each blank with the correct answer.

Male reproductive system check-up

The penile shaft consists of three columns of _____ bound together
by _____ . Two _____ form the major part of the
penis. On the underside, the _____ encases the urethra. The
_____ , at the distal end of the shaft, is a cone-shaped structure
formed from the corpus spongiosum; its lateral margin forms a ridge of tissue
known as the _____ . The _____ opens through the
glans to allow urination and ejaculation. The penis receives blood through the
_____ artery. Venous blood returns through the _____
to the vena cava.

Options
corona
corpora cavernosa
corpus spongiosum
erectile tissue
glans penis
heavy fibrous tissue
internal iliac vein
internal pudendal
urethral meatus

■ Match point

Match each term on the left with its definition on the right.

1. Spermatic cord _____
2. Tunica vaginalis _____
3. Tunica albuginea _____
4. Seminiferous tubules _____
5. Dartos muscle _____
6. Cremaster muscle _____

A. Elevates the testes, which helps govern temperature of the testes

B. Outer layer of connective tissue enveloping the testes

C. Connective tissue sheath that encases autonomic nerve fibers, blood and lymph vessels, and the vas deferens

D. Causes scrotal skin to wrinkle, which helps regulate temperature

E. Small tubes in which spermatogenesis takes place

F. Inner layer of connective tissue enveloping the testes

An important function of the scrotum is to keep the testes cooler than the rest of the body.

Train your brain

Sound out the pictures and symbols to discover a fact about the anatomy of the male reproductive system.

Jumble gym

Use the clues to help you unscramble words related to the male reproductive duct system.
Then use the circled letters to answer the question posed.

Question: **What structure divides the scrotum into two parts?**

1. Coiled tube located superior to and along the posterior border of the testis

   D I D P I E I S M Y   _ _ _ _ _ _ _ (_) _ _

2. Leads from the testes to the abdominal cavity, extends through the inguinal canal, arches over the urethra, and descends behind the bladder

   S V A   E N D F R E E S   _ _ _ _(_)_ _ _ _ _ _(_)

3. Enlarged portion of the structure named in the previous question, which merges with the duct of the seminal vesicle to form the short ejaculatory duct

   L L A M A U P   _ _(_)_ _ _ _

4. Small tube leading from the floor of the bladder to the exterior

   R E A R H U T   (_)_ _(_)_ _ _

Answer: _ _ _ _ _ _

Power stretch

Unscramble the words below to discover the names of the male accessory reproductive glands.
Then draw lines linking each gland to its particular characteristics.

MEALSIN SECLIVES

— — — — — — —
— — — — — — — —

ARUBBLEHURLTO SNAGLD

— — — — — — — — — — — — — —
— — — — — —

ASPOTTER DALNG

— — — — — — — —
— — — — —

A. Lies under the bladder and surrounds the urethra

B. Produce about 60% of the fluid portion of semen

C. Continuously secretes a thin, milky, alkaline fluid that enhances sperm motility

D. Paired glands located inferior to the prostate

E. Paired sacs at the base of the bladder

F. Secretes a viscid fluid that becomes about 10% of the fluid portion of semen

G. Produces about 30% of the fluid portion of semen

Circuit training

Trace the stages of spermatogenesis as well as the path mature sperm take by drawing arrows between the boxes below.

| Spermatozoa | Seminiferous tubules |

| Spermatogonia | Divide |

| Spermatid | Epididymis |

| Primary spermatocytes | Divide |

| Secondary spermatocytes | Vas deferens |

Strike out

Some of the following statements about sperm are incorrect. Cross out all of the incorrect statements.

1. Sperm formation begins at birth, although sperm don't mature until a male reaches puberty.

2. Primary germinal epithelial cells are called spermatogonia.

3. Both spermatogonia and primary spermatocytes contain 46 sex chromosomes.

4. When primary spermatocytes divide, no new chromosomes are formed.

5. Each secondary spermatocyte contains 22 autosomes.

6. Each secondary spermatocyte also contains both an X and a Y chromosome.

7. The head of the sperm contains the nucleus.

8. The tail of the sperm contains a large amount of adenosine triphosphate, which provides energy for sperm motility.

9. Newly mature sperm are stored in the epididymis until emission.

10. Sperm cells lose their potency after 4 days.

Number and motility affect fertility. A low sperm count (less than 20 million per milliliter of ejaculated semen) may cause infertility.

■■
■ Hit or miss

Some of the following statements are true; the others are false. Mark each accordingly.

_____ 1. Androgens are produced in the pituitary gland.

_____ 2. The adrenal glands secrete testosterone.

_____ 3. Testosterone is responsible for development and maintenance of male sex organs and secondary sex characteristics.

_____ 4. Testosterone is required for spermatogenesis.

_____ 5. Testosterone secretion begins at the onset of puberty.

_____ 6. The presence or absence of testosterone in utero determines sexual differentiation of the fetus.

_____ 7. The pituitary gland begins secreting gonadotropins at the onset of puberty.

_____ 8. A male usually achieves full physical maturity by age 16.

_____ 9. Male sexual and reproductive function remains fairly consistent throughout life.

_____ 10. With age, the prostate gland shrinks and its secretions diminish.

Coaching session
Male reproductive changes with aging

- Reduced testosterone production
  - Decreased libido
  - Testicular atrophy
  - Decreased sperm production (by as much as 69% between ages 60 to 80)
- Enlargement of prostate gland
- Decreased seminal fluid

Match point

Match the following terms associated with the female reproductive system with their definitions.

1. Mons pubis _____
2. Labia majora _____
3. Menarche _____
4. Labia minora _____
5. Prepuce _____
6. Frenulum _____
7. Fourchette _____
8. Clitoris _____
9. Vestibule _____
10. Skene's glands _____
11. Bartholin's glands _____
12. Hymen _____

A. The onset of menses

B. Small, protuberant organ just beneath the arch of the mons pubis

C. Tissue membrane covering the vaginal orifice

D. Rounded cushion of fatty connective tissue over the symphysis pubis

E. Oval area bounded anteriorly by the clitoris, laterally by the labia minora, and posteriorly by the fourchette

F. Posterior portion of the clitoris

G. Two raised folds of adipose and connective tissue that border either side of the vulva, extending from the mons pubis to the perineum

H. Two folds of mucosal tissue that lie within and alongside the labia majora

I. Mucus-producing glands found on both sides of the urethral opening

J. Thin tissue fold along the anterior edge of the perineum

K. Located laterally and posteriorly on either side of the inner vaginal orifice

L. Hoodlike covering over the clitoris

Finish line

Identify the structures of the female external genitalia shown in the illustration below.

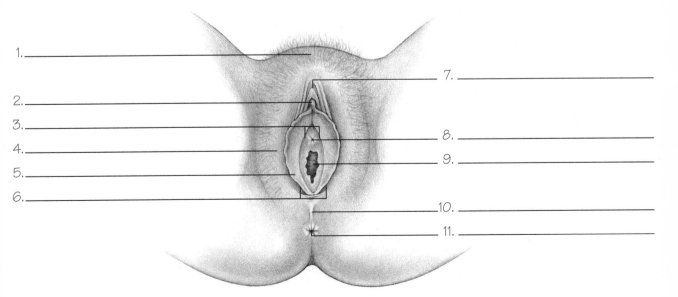

1. _____
2. _____
3. _____
4. _____
5. _____
6. _____
7. _____
8. _____
9. _____
10. _____
11. _____

Finish line

The following illustration shows a lateral view of the female internal genitalia. See if you can identify all the structures.

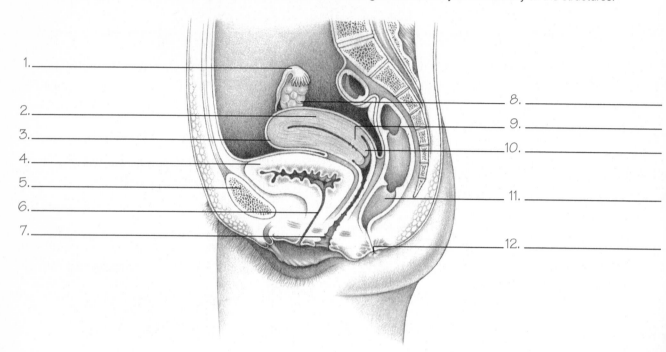

1. _____

2. _____

3. _____

4. _____

5. _____

6. _____

7. _____

8. _____

9. _____

10. _____

11. _____

12. _____

Pep talk

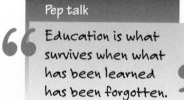

"Education is what survives when what has been learned has been forgotten.

—B.F. Skinner

■■
■ Hit or miss

Some of the following statements about the female genitalia are true; the others are false. Mark each accordingly.

_____ 1. The vagina is a highly elastic tube composed entirely of muscle tissue.

_____ 2. Surrounding the cervix are four recesses in the vaginal wall, called fornices.

_____ 3. The perineum connects the uterus to the vaginal vault.

_____ 4. The labia majora contain sebaceous glands that secrete a bactericidal lubricant.

_____ 5. The labia are highly vascular and have many nerve endings, making them sensitive to pain, pressure, touch, and sexual stimulation.

_____ 6. The vagina's only function is to accommodate the penis during coitus.

_____ 7. The cervix is highly elastic and resumes its normal shape within a few weeks of childbirth.

■■
■ Match point

Match each structure listed on the left with its location and functions, listed on the right.

1. Urethral meatus _____

2. Inferior vesical arteries _____

3. Internal pudendal arteries _____

4. Branches of the uterine arteries _____

5. External os _____

6. Internal os _____

7. Fundus _____

8. Corpus _____

A. Arteries that supply blood to the upper vagina

B. Main uterine body

C. Arteries that supply blood to the middle vagina

D. Slitlike opening below the clitoris through which urine leaves the body

E. Upper portion of the uterus

F. Lower cervical opening

G. Arteries that supply blood to the lower vagina

H. Upper cervical opening

Jumble gym

Use the clues to help you unscramble words related to the female reproductive system. Then use the circled letters to answer the question posed.

Question: In what location does fertilization take place?

1. Portion of the uterus that accommodates most of the growing fetus until term

   D U N F U S Ⓞ _ _ _ Ⓞ _

2. The funnel-shaped ending of the fallopian tube

   B U I L D F U N M U _ _ _ _ _ Ⓞ _ Ⓞ _ _

3. Blood returning from the vagina empties into these veins

   A C O P Y R I G H T S _ _ Ⓞ _ _ Ⓞ _ Ⓞ _ _ _

4. Fingerlike projections at the end of the fallopian tubes

   M A B I F I R E _ Ⓞ _ _ _ _ _ Ⓞ

5. Found in the ovaries at birth

   A G A I N F A R   C L O S E F I L L

   _ _ Ⓞ _ _ _ _ Ⓞ   _ Ⓞ _ _ _ _ Ⓞ _ _

Answer: _ _ _ _ _ _ _ _   _ _ _ _

Strike out

Some of the following statements about the ovaries and ova are incorrect. Cross out all of the incorrect statements.

1. The size, shape, and position of the ovaries vary with age.

2. Ovaries are round, smooth, and pink at puberty.

3. At menopause, ovaries take on an almond shape and a rough, pitted surface.

4. At puberty, ovaries begin to produce graafian follicles.

5. During the childbearing years, one graafian follicle produces a mature ovum during the first half of each menstrual cycle.

6. As the ovum matures, the follicle ruptures and the ovum is swept into the fallopian tube.

7. The ovaries' only function is to produce mature ova.

Batter's box

Fill in the blanks with the appropriate words.

Keeping abreast

The _____ glands, which are located in the breasts, are specialized
_____ (1)

_____ glands that secrete _____ . Each gland contains
(2)   (3)

15 to 25 _____ separated by fibrous connective tissue and fat. Tiny,
(4)

saclike duct terminals called _____ secrete milk during lactation.
(5)

The ducts draining the lobules converge to form _____ , or
(6)

excretory, ducts. Sinuses called _____ store milk during lactation.
(7)

The pigmented area in the center of the breast is called the _____ .
(8)

Fibrous _____ support the breasts. Sebaceous glands on the areolar
(9)

surface called _____ produce _____ , which
(10)   (11)

lubricates the areolae and nipples during breast-feeding.

Options
accessory
acini
ampullae
areola
Coopers' ligaments
lactiferous
lobes
mammary
milk
Montgomery's tubercles
sebum

Hit or miss

Some of the following statements about the female reproductive cycle are true; the others are false. Mark each accordingly.

_____ 1. The female reproductive cycle usually lasts 38 days.

_____ 2. During this cycle, ovulatory, hormonal, and endometrial changes occur sequentially.

_____ 3. Ovulatory changes begin on the first day of the menstrual cycle.

_____ 4. The luteal phase is the phase of the cycle in which a follicle develops.

_____ 5. After ovulation, the collapsed follicle forms the corpus luteum.

_____ 6. If fertilization doesn't occur, the corpus luteum degenerates.

Finish line

The following illustration shows a lateral cross section of a female breast. See if you can identify the structures shown.

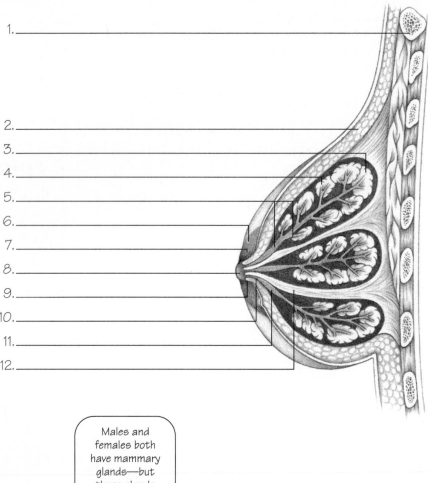

1._____

2._____

3._____

4._____

5._____

6._____

7._____

8._____

9._____

10._____

11._____

12._____

Males and females both have mammary glands—but these glands function only in females.

▪▪
▪ Circuit training

Trace the sequence of events leading to ovulation by drawing arrows between the boxes below.

Stimulates anterior pituitary to secrete FSH and LH

Spur follicle to develop

Low estrogen and progesterone levels

Stimulate hypothalamus to secrete Gn-RH

Ovulation

Triggers folllicular rupture

LH level spikes

▪▪
▪ Batter's box

Fill in each blank with the correct answer. *Hint:* Some answers may be used more than once.

Getting into the endometrium

The endometrium is most receptive to implantation about _____ days after

the initiation of ovulation. In the first _____ days of the reproductive cycle,
 1 2

the endometrium sheds its _____ layer, leaving the _____
 3 4

layer (deepest layer) intact. The endometrium begins regenerating its functional layer at

about day 6, (the _____ phase), spurred by rising _____
 5 6

levels. After ovulation (about day _____ , increased _____
 7 8

secretion stimulates the functional layer into a secretory mucosa (_____
 9

phase), which is more receptive to implantation of the fertilized ovum. If implantation

doesn't occur, the _____ degenerates, _____ levels drop, and
 10 11

the endometrium again sheds its functional layer.

Options
7
5
14
basal
corpus luteum
estrogen
functional
progesterone
proliferative
secretory

Strike out

Some of the following statements about female reproductive changes with aging are incorrect. Cross out all of the incorrect statements.

1. Ovulation continues until the time of menopause.

2. As the ovaries reach the end of their productive cycle, they become unresponsive to estrogen stimulation.

3. Cessation of menstruation usually occurs between ages 40 and 55.

4. Vulval tissue shrinks with age, exposing the sensitive area around the urethra and vagina to abrasions and irritations.

5. The pH of vaginal secretions decreases, making the risk of vaginal infections minimal.

6. Symptoms of relaxation of pelvic support (common in postreproductive women) include lower backaches, a feeling of pelvic heaviness, and difficulty in rising from a sitting position.

Train your brain

Sound out the pictures and symbols to reveal information about a change that occurs in the female reproductive system with age.

Reproduction and lactation

Reproduction and lactation review

Fertilization

- Union of a spermatozoon and an ovum to form a single cell
- After, cells of fertilized ovum begin dividing as the ovum travels to the uterine cavity, where it implants in the uterine lining

Cervical mucus

- Protects spermatozoa from acidic vaginal secretions
- Penetrability by spermatozoa depends on the phase of the menstrual cycle at the time of transit
  - During midcycle, mucus is relatively thin and spermatozoa can pass readily through

Sperm

- Can survive in the reproductive tract for up to 4 days
- Are typically able to fertilize the ovum for up to 2 days after ejaculation

Steps in fertilization

- Sperm penetrates cervical mucus
- Sperm disperse granulosa cells and penetrate the zona pellucida that surround the incompletely developed ovum
- After penetration, the zona pellucida prevents penetration by other sperm
- Spermatozoon's head fuses with the ovum nucleus, creating a cell nucleus with 46 chromosomes
  - The fertilized ovum is called a *zygote*
- Cells of zygote begin to divide
- Zygote travels to the uterine cavity and implants in the uterine lining

Pregnancy

- Starts with fertilization and ends with childbirth
- Average duration (called *gestation*) is 38 weeks
- Complex sequence of preembryonic, embryonic, and fetal development transforms the zygote into a full-term fetus
- Expected delivery date (EDD) is typically calculated from the beginning of the last menstrual period (LMP)
- Nägele's rule to calculate EDD: Count back 3 months from the first day of the LMP and then add 7 days

Stages of fetal development

PREEMBRYONIC PERIOD (FERTILIZATION TO WEEK 3)

- Starts with fertilization and lasts 3 weeks
- Zygote passes through the fallopian tube and undergoes a series of mitotic divisions, or cleavage

EMBRYONIC PERIOD (WEEKS 4 THROUGH 7)

- Developing zygote starts to take on a human shape
  - Now called an *embryo*
- Each germ layer (ectoderm, mesoderm, and endoderm) starts to form specific tissues in the embryo
- Organ systems form
- Embryo is particularly vulnerable to injury, such as by maternal drug use, certain maternal infections, and other factors

FETAL PERIOD (WEEK 8 THROUGH BIRTH)

- Maturing fetus enlarges and grows heavier
- Characterized by two unusual features:
  - Fetal head is disproportionately large for its body
  - Fetus lacks subcutaneous fat

Structural changes in the ovaries and uterus

CORPUS LUTEUM

- Normal functioning requires continual stimulation by luteinizing hormone (LH)
- Progesterone produced by the corpus luteum suppresses LH release by the pituitary gland
- If pregnancy occurs, the corpus luteum continues to produce progesterone until the placenta takes over
- If pregnancy doesn't occur, the corpus luteum atrophies 3 days before menstrual flow begins
- Pregnancy stimulates the placental tissue to secrete large amounts of human chorionic gonadotropin (HCG), which prevents corpus luteum degeneration and stimulates the corpus luteum to produce large amounts of estrogen and progesterone
- The corpus luteum, stimulated by the hormone HCG, produces the estrogen and progesterone needed to maintain the pregnancy during the first 3 months
- HCG level gradually increases during this time, peaks at about 10 weeks' gestation, and then gradually declines

DECIDUA

- Endometrial lining of the uterus that undergoes the hormone-induced changes of pregnancy
- Decidual cells secrete three substances:
  - Prolactin—promotes lactation

– Relaxin—induces relaxation of the connective tissue of the symphysis pubis and pelvic ligaments and promotes cervical dilation

– Prostaglandin

Amniotic sac

■ Gradually increases in size and surrounds the embryo, eventually filling the cavity and fusing with the chorion by the eighth week of gestation

Amniotic fluid

■ Gives the fetus a buoyant, temperature-controlled environment

■ Serves as a fluid wedge that helps open the cervix during birth

■ Early in pregnancy, comes from two primary sources:

– Fluid filtering into the amniotic sac from maternal blood as it passes through the uterus or from fetal blood passing through the placenta

– Fluid diffusing into the amniotic sac from the fetal skin and respiratory tract

■ Later, when the fetal kidneys begin to function, major source of amniotic fluid is fetal urine

Yolk sac

■ Forms next to the endoderm of the germ disk

■ A portion is incorporated in the developing embryo and forms the GI tract

■ Another portion develops into primitive germ cells, which travel to the developing gonads and eventually form oocytes (the precursor of the ovum) or spermatocytes (the precursor of the spermatozoon) after gender has been determined

■ During early embryonic development, also forms blood cells

■ Eventually, atrophies and disintegrates

Placenta

■ Provides nutrients to and removes wastes from the fetus between the third month of pregnancy and birth

■ Formed from the chorion, its chorionic villi, and the adjacent decidua basalis

■ Attached to fetus by the umbilical cord

– Two umbilical arteries transport blood from the fetus to the placenta

– Placental veins gather blood returning from the villi and join to form a single umbilical vein, which enters the cord and returns blood to the fetus

– Uteroplacental circulation—carries oxygenated arterial blood from the maternal circulation to the intervillous spaces and leaves the intervillous spaces and flows back into the maternal circulation through veins in the basal part of the placenta near the arteries

– Fetoplacental circulation—transports oxygen depleted blood from the fetus to the chorionic villi by the umbilical arteries and returns oxygenated blood to the fetus through the umbilical vein

■ By end of third month, produces most of the hormones

– Estrogen—stimulates uterine development to provide a suitable environment for the fetus

Progesterone—synthesized by the placenta from maternal cholesterol; reduces uterine muscle irritability and prevents spontaneous abortion of the fetus

– Human placental lactogen—stimulates maternal protein and fat metabolism to ensure a sufficient supply of amino acids and fatty acids for the mother and fetus; also stimulates breast growth in preparation for lactation

Labor and the postpartum period

■ Childbirth (delivery of the fetus) is achieved through labor (the process in which uterine contractions expel the fetus from the uterus)

– When labor begins, contractions become strong and regular

– Eventually, voluntary bearing-down efforts supplement the contractions, resulting in delivery of the fetus and placenta

Onset of labor

■ The number of oxytocin receptors on uterine muscle fibers increase progressively during pregnancy, peaking just before labor onset and makes the uterus more sensitive to the effects of oxytocin

■ Stretching of the uterus over the course of the pregnancy initiates nerve impulses that stimulate oxytocin secretion from the posterior pituitary lobe

■ Near term, the fetal pituitary gland secretes more adrenocorticotropic hormone, which causes the fetal adrenal glands to secrete more cortisol

■ The cortisol heightens oxytocin and estrogen secretion, and reduces progesterone secretion which intensifies uterine muscle irritability and makes the uterus more sensitive to oxytocin stimulation

■ Declining progesterone levels convert esterified arachidonic acid to form prostaglandins, which, in turn, diffuse into the uterine myometrium, thereby inducing uterine contractions

■ Oxytocin secretions may also stimulate prostaglandin formation by the decidua

■ As the cervix dilates, nerve impulses are transmitted to the central nervous system, causing an increase in oxytocin secretion from the pituitary gland

■ Acting as a positive feedback mechanism, increased oxytocin secretion stimulates more uterine contractions, which further dilate the cervix and lead the pituitary to secrete more oxytocin

Stages of childbirth

- Duration of each varies with the size of the uterus, the woman's age, and the number of previous pregnancies

FIRST STAGE

- Fetus begins its descent
- Marked by cervical effacement (thinning) and dilation
- Full cervical dilatation by end of stage
- Lasts 6 to 24 hours in primiparous women
- Is often significantly shorter for multiparous women

SECOND STAGE

- Begins with full cervical dilation
- Ends with delivery of the fetus
- Amniotic sac ruptures
- Uterine contractions increase in frequency and intensity
- Maternal muscles force the flexed head of the fetus to rotate anteriorly and cause the back of the head to move under the symphysis pubis
- As the uterus contracts, the flexed head of the fetus is forced deeper into the pelvis
- Resistance of the pelvic floor gradually forces the head to extend
- As the head presses against the pelvic floor, vulvar tissues stretch and the anus dilates
- The head of the fetus rotates back to its former position after passing through the vulvovaginal orifice (usually, head rotation is lateral [external] as the anterior shoulder rotates forward to pass under the pubic arch)
- Delivery of the shoulders and the rest of the fetus follows
- Averages about 45 minutes in primiparous women; may be much shorter in multiparous women

THIRD STAGE

- Starts immediately after childbirth and ends with placenta expulsion
- After fetus is delivered, the uterus continues to contract intermittently and grows smaller
- Placenta, which can't decrease in size, separates from the uterus, and blood seeps into the area of placental separation
- Averages about 10 minutes in both primiparous and multiparous women

Postpartum period

- After childbirth, the reproductive tract takes about 6 weeks to revert to its former condition in a process called *involution*
- Uterus quickly grows smaller, with most involution taking place during the first 2 weeks after delivery

LOCHIA

- Lochia rubra—a bloody discharge, appears 1 to 4 days postpartum
- Lochia serosa—a pinkish brown, serous discharge that occurs from 5 to 7 days postpartum
- Lochia alba—a grayish white or colorless discharge that occurs from 1 to 3 weeks postpartum

Lactation

- Governed by interactions involving four hormones:
  - Estrogen and progesterone—produced by the ovaries and placenta
  - Prolactin and oxytocin—produced by the pituitary gland under hypothalamic control

Hormonal initiation of lactation

- Placental production of estrogen and progesterone increases, causing glandular and ductal tissue in the breasts to proliferate
- Prolactin causes milk secretion after breast stimulation by estrogen and progesterone
- Oxytocin from the posterior pituitary lobe causes contraction of specialized cells in the breast, producing a squeezing effect that forces milk down the ducts
- Breast-feeding, in turn, stimulates prolactin secretion, resulting in a high prolactin level that induces changes in the menstrual cycle
- Progesterone and estrogen levels fall after delivery
- Estrogen and progesterone no longer inhibiting prolactin's effects on milk production and the mammary glands start to secrete milk
- Nipple stimulation during breast-feeding results in transmission of sensory impulses from the nipples to the hypothalamus
- If the nipples aren't stimulated by breast-feeding, prolactin secretion declines after delivery
- Milk secretion continues as long as breast-feeding regularly stimulates the nipples
- If breast-feeding stops, the stimulus for prolactin release is eliminated and milk production ceases

Breast-feeding and the menstrual cycle

- During the postpartum period, a woman's high prolactin level inhibits FSH and LH release
- Prolactin output soon drops, ending inhibition of FSH and LH production by the pituitary in the absence of breast-feeding, resulting in cyclic release of FSH and LH
- In a breast-feeding woman, the menstrual cycle doesn't resume because prolactin inhibits the cyclic release of FSH and LH necessary for ovulation
- Prolactin release in response to breast-feeding gradually declines, as does the inhibitory effect of prolactin on FSH and LH release
- Ovulation and the menstrual cycle may resume and pregnancy may occur after this, even if the woman continues to breast-feed

▪▪ Batter's box

Fill in the blanks with the appropriate words.

Joining forces

Creation of a new human being begins with _____ , union of a _____
₁ ₂

and an _____ to form a single cell. First, however, a spermatozoon must survive
₃

the acidic secretions of the _____ . Only the spermatozoa that survive enter the
₄

cervical _____ , where they're protected by cervical _____ .
₅ ₆

The ability of the spermatozoa to penetrate the cervical mucus depends on the phase of the

_____ cycle at the time of transit. After fertilization, the cells of the fertilized ovum
₇

begin dividing as the ovum travels to the _____ , where it implants in the
₈

_____ .
₉

The fruits of labor

During a 38-week period of _____ , a complex sequence of development transforms
₁₀

the _____ into a full-term _____ . Childbirth is achieved through
₁₁ ₁₂

_____ , the process by which uterine contractions expel the fetus from the uterus.
₁₃

Options
canal
endometrium
fertilization
fetus
gestation
labor
menstrual
mucus
ovum
spermatozoon
uterine cavity
vagina
zygote

Power stretch

Unscramble the following words to discover the names of the three major stages of development during pregnancy. Then draw lines to link each stage with its particular characteristics.

REP-CRIBMONEY

— — — –

— — — — — — — — —

BIMERCYNO

— — — — — — — — —

ALEFT

— — — — —

A. Phase beginning at the ninth week of gestation and lasting until birth

B. Phase in which the zygote develops into a small mass of cells called a morula

C. Phase in which the zygote starts to take on a human shape

D. Phase beginning with ovum fertilization and lasting for 2 weeks

E. Phase in which the fetus lacks subcutaneous fat

F. Phase in which the blastocyst attaches to the endometrium

G. Phase beginning at the third week of gestation and lasting through the eighth week

H. Phase in which the zygote is called an embryo

I. Phase in which the fetus's head is disproportionately large compared with its body

J. Phase in which the organ systems form

K. Phase in which the zygote goes through a series of mitotic divisions

You make the call

Using the following illustrations as a guide, describe how fertilization occurs.

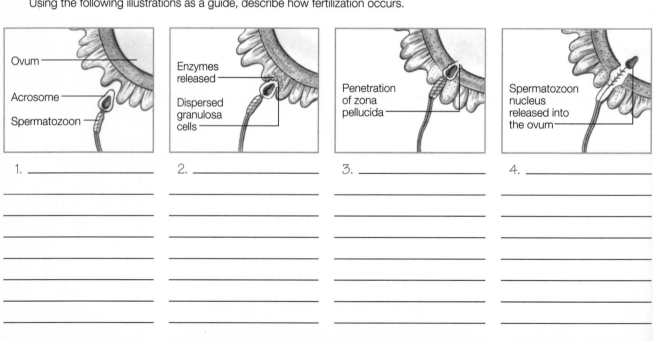

Ovum

Acrosome

Spermatozoon

1. _____

Enzymes released

Dispersed granulosa cells

2. _____

Penetration of zona pellucida

3. _____

Spermatozoon nucleus released into the ovum

4. _____

Cross-training

Test your knowledge of terms related to reproduction and lactation by completing the following crossword puzzle.

Across

4. Endometrial lining that undergoes the hormone-induced changes of pregnancy

5. Initial breast milk that's yellow in color

7. Fertilized ovum

12. Progressive enlargement of the cervical os during labor

14. Head cap of the spermatozoon

Down

1. Precursors to spermatozoa

2. Hormone that promotes lactation

3. Milk synthesis and secretion by the breasts

6. Outermost germ layer

8. Pituitary hormone that stimulates uterine contractions

9. Hormone that induces relaxation of the connective tissue of the symphysis pubis and pelvic ligaments and promotes cervical dilation

10. Innermost germ layer

11. Middle germ layer

13. Precursors to ova

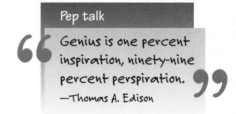

Pep talk

"Genius is one percent inspiration, ninety-nine percent perspiration."
—Thomas A. Edison

Starting lineup

Test your knowledge of the stages of normal childbirth by putting the following steps in the correct order.

Placenta separates from uterus.
Fetus begins descent.
Uterus continues to contract and shrinks.
Placenta is delivered.
Fetus is delivered.
Full cervical dilation occurs.
Amniotic sac ruptures.
Contractions increase in frequency and intensity.
Fetus undergoes cardinal movements of labor.
Fetal head is delivered.
Contractions cause cervical effacement.
Fetal shoulders are delivered.

First stage

1.

2.

3.

Second stage

4.

5.

6.

7.

8.

9.

Third stage

10.

11.

12.

Answers

Chapter 1

Page 4

Cross-training

Crossword puzzle answers:
1. ABDOMINAL
3. PELVIC
4. DIGITAL
5. DORSAL
6. TRANSVERSE
7. SUPERFICIAL
8. FRONTAL
9. THORACIC
10. PROXIMAL
11. INFERIOR
12. VENTRAL
13. CRANIAL
14. OBLIQUE
15. SUPERIOR
16. PERITONEUM
17. MEDIAL
18. LATERAL

Page 5

Finish line

1. Cytoplasm
2. Lysosome
3. Mitochondrion
4. Nucleus
5. Golgi apparatus
6. Ribosomes
7. Endoplasmic reticulum
8. Cell membrane

Match point

1. E, 2. K, 3. C, 4. F, 5. J, 6. G, 7. D, 8. A, 9. I, 10. B, 11. L, 12. H

Page 6

Hit or miss

1. True.
2. False. Meiosis consists of two divisions. The first results in two daughter cells, each containing 23 chromosomes. The second results in two more daughter cells (or four total), each containing 23 chromosomes.
3. True.

You make the call

1. In diffusion, *solutes* move from an area of *higher* concentration to one of *lower* concentration. Eventually an equal distribution of solutes occurs.
2. In osmosis, *fluid* moves passively across a membrane, from an area of *lower* solute concentration (comparatively more fluid) into an area of *higher* solute concentration (comparatively less fluid).
3. In active transport, a *substance* moves across the cell membrane from an area of *lower* concentration to one of *higher* concentration. This requires energy.

Chapter 2

Page 10

Boxing match

1. Genotype, 2. Genetics, 3. Phenotype, 4. Gamete,
5. Chromosome, 6. Genome, 7. Locus, 8. Mutation,
9. Meiosis, 10. Alleles

Hit or miss

1. False. A human ovum contains 23 chromosomes. A sperm also contains 23 chromosomes. When an ovum and a sperm unite, the corresponding chromosomes pair up, resulting in 46 chromosomes, or 23 pairs.
2. True.
3. False. The two sex chromosomes on the 23rd pair determine gender; the other 22 pairs have no effect on gender.
4. True.
5. True.
6. False. Each gamete produced by a male contains either an X or a Y chromosome.
7. False. Each gamete produced by a female contains only an X chromosome. When a sperm with an X chromosome fertilizes an ovum, the offspring is female (two X chromosomes).
8. True.
9. False. The location, or locus, of each gene is specific and doesn't vary from person to person. This allows the genes in an ovum to join the corresponding genes from a sperm when the chromosomes pair up at fertilization.

Page 11

Strike out

2. ~~Recessive genes are expressed only under the influence of environmental factors.~~ Recessive genes are expressed when both parents transmit it to the offspring.
3. ~~Males have more genetic material than females.~~ Because females carry two X chromosomes, they carry more genetic material than males, who carry one X and one Y chromosome.
6. ~~Because sex-linked genes are recessive, they are rarely expressed.~~ In males, sex-linked genes behave like dominant genes because no second X chromosome exists.

Parallel bars

1. Ova, 2. XY, 3. Phenotype, 4. Heterozygous, 5. Dominant,
6. Monosomy

Page 12

Jumble gym

1. Mutation, 2. Multifactorial, 3. Congenital anomalies,
4. Translocation, 5. Monosomy, 6. Trisomy,
7. Nondisjunction

Answer: Autosomal

Page 13

Cross-training

					1 E	L	E	2 M	E	N	T			3 V
					L			A						A
	4 N		5 H	O	M	E	O	S	T	A	S	6 I	S	L
	E		C					T			N			E
	U		7 I	S	O	T	O	P	E		O			N
	T		R			R		R			R			T
	R		8 M	O				9 O	R	G	A	N	I	C
10 C	O	M	P	O	U	N	D			A			E
	N		L		S	11 P				N			
	S		E		12 P	R	O	T	E	I	N		
			C		O			C					
13 N	U	C	L	E	U	S	14 A	T	O	M			
			L		O								
			E		15 E	N	E	R	G	Y			
					S								

Page 14

Match point

1. E, 2. C, 3. H, 4. D, 5. F, 6. G, 7. B, 8. J, 9. K, 10. I, 11. M, 12. L, 13. A

You make the call

1. A synthesis reaction combines two or more substances (reactants) to form a new, more complex substance (product). This results in a chemical bond.
2. In a decomposition reaction, a substance decomposes, or breaks down, into two or more simpler substances, leading to the breakdown of a chemical bond.
3. An exchange reaction is a combination of a decomposition and a synthesis reaction. This reaction occurs when two complex substances decompose into simpler substances. The simple substances then join (through synthesis) with different simple substances to form new complex substances.
4. In a reversible reaction, the product reverts to its original reactants, and vice versa. Reversible reactions may require special conditions, such as light or heat.

Chapter 3

■ Page 18
Cross-training

The completed crossword:

1 (across) LUNULA
5 (across) DERMIS
6 (across) LANGERHANS
11 (across) DERMATOMES
13 (across) EPIDERMIS

Down answers: APOCRINE (2), MELANOCYTES (3), SEBACEOUS (4), LIPIDS (5), COLLAGEN (8), ECCRINE (9), KERATIN... DERMIS (7), MATRIX (10), ELASTIN (12)

■ Page 19
Power stretch

Protection: A, D, E, G, H, J
Sensory perception: B, I
Body temperature regulation: F, I
Excretion: C, F

■ Page 20
A-maze-ing race

1. B. When internal body temperature rises, small arteries in the skin dilate.
3. A. Arterial dilation increases blood flow.
6. B. Sweat evaporation cools the skin.

■ Page 21
Match point

1. F, 2. C, 3. G, 4. A, 5. H, 6. E, 7. B, 8. I, 9. D, 10. J

Finish line

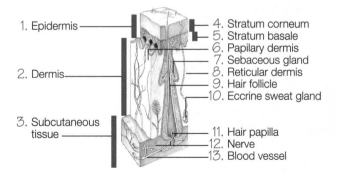

1. Epidermis
2. Dermis
3. Subcutaneous tissue
4. Stratum corneum
5. Stratum basale
6. Papilary dermis
7. Sebaceous gland
8. Reticular dermis
9. Hair follicle
10. Eccrine sweat gland
11. Hair papilla
12. Nerve
13. Blood vessel

■ Page 22

Hit or miss

1. False. Hairs are composed of keratin.
2. True.
3. False. Nails are composed of a specialized type of keratin.
4. True.
5. False. Sebaceous glands occur on all parts of the skin except the palms and soles.
6. True.
7. True.
8. False. Hair follicles have a rich blood and nerve supply.
9. True.
10. True.
11. False. Apocrine glands begin to function at puberty.

Finish line

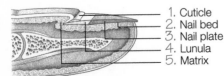

1. Cuticle
2. Nail bed
3. Nail plate
4. Lunula
5. Matrix

■ Chapter 4

■ Page 27

Power stretch

Cardiac: D, F
Smooth: C
Skeletal: A, B, E

Hit or miss

1. True.
2. False. Most movement involves groups of muscles rather than one muscle.
3. True.
4. True.
5. False. Skeletal muscle contains transverse stripes, called *striations.*
6. True.
7. False. The epimysium is stronger than the perimysium; it binds all of the fasciculi together to form the entire muscle.
8. False. The epimysium becomes a tendon.
9. True.
10. True.

■ Page 28

Cross-training

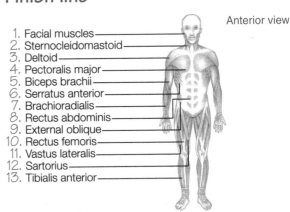

■ Page 29

Finish line

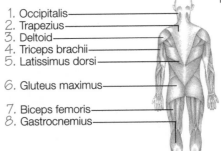

Anterior view

1. Facial muscles
2. Sternocleidomastoid
3. Deltoid
4. Pectoralis major
5. Biceps brachii
6. Serratus anterior
7. Brachioradialis
8. Rectus abdominis
9. External oblique
10. Rectus femoris
11. Vastus lateralis
12. Sartorius
13. Tibialis anterior

Posterior view

1. Occipitalis
2. Trapezius
3. Deltoid
4. Triceps brachii
5. Latissimus dorsi
6. Gluteus maximus
7. Biceps femoris
8. Gastrocnemius

Page 30

You make the call

1. Retraction, 2. Protraction, 3. Circumduction, 4. Extension,
5. Flexion, 6. Pronation, 7. Supination, 8. Eversion,
9. Inversion, 10. Abduction, 11. Adduction, 12. Internal
rotation, 13. External rotation

Page 31

Finish line

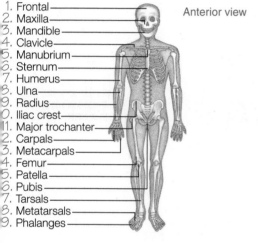

Anterior view

1. Frontal
2. Maxilla
3. Mandible
4. Clavicle
5. Manubrium
6. Sternum
7. Humerus
8. Ulna
9. Radius
10. Iliac crest
11. Major trochanter
12. Carpals
13. Metacarpals
14. Femur
15. Patella
16. Pubis
17. Tarsals
18. Metatarsals
19. Phalanges

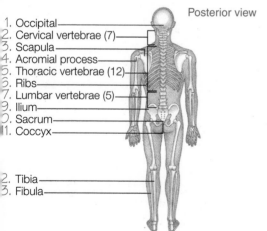

Posterior view

1. Occipital
2. Cervical vertebrae (7)
3. Scapula
4. Acromial process
5. Thoracic vertebrae (12)
6. Ribs
7. Lumbar vertebrae (5)
9. Ilium
10. Sacrum
11. Coccyx

12. Tibia
13. Fibula

Page 32

Match point

1. B, 2. G, 3. E, 4. J, 5. I, 6. A, 7. D, 8. H, 9. F, 10. C

Power stretch

Axial: A, D, E, G, J, K, M , P
Appendicular: B, C, F, H, I, L, N, O

Page 33

Odd man out

1. Scapula (long bones), 2. Radius (short bones), 3. Ulna (flat
bones), 4. Femur (irregular bones). The fifth bone type is
sesamoid, of which the patella is an example.

Strike out

2. ~~Lengthening of arms.~~ Arms don't lengthen, although they may
appear long in relation to a shortened trunk.
4. ~~Accelerated collagen formation.~~ Collagen formation declines.
5. ~~Increased red blood cell production.~~ A change in red blood
cell production isn't associated with normal aging.
7. ~~Thinning of synovial membranes.~~ Synovial membranes
become more fibrotic.

Memory jogger

Movement; **P**osture; **H**eat

Page 34

Finish line

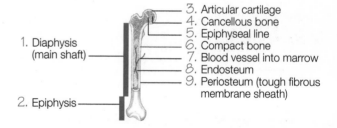

1. Diaphysis
(main shaft)

2. Epiphysis

3. Articular cartilage
4. Cancellous bone
5. Epiphyseal line
6. Compact bone
7. Blood vessel into marrow
8. Endosteum
9. Periosteum (tough fibrous
membrane sheath)

Page 35

Train your brain

Bones produce blood cells and store mineral salts.

Power stretch

Compact: B, E, G H
Cancellous: A, C, D, F

■ Page 36

Cross-training

```
 1              2
 L  A  M  E  L  L  A  E
 A        P
 C        I     3 C              4
 U        P       O              O
 N       5 H  E  M  A  T 6 O  P  O  I  E  S  I  S     7 C
 A        Y          S          I              A
 E        S          T    8 C   F              N
          I          E      A   I      9 D     C
          S          O      R   C        I     E
                 10 O B      T   A        A     L
                    S  L     I   T        P     L
       11 J  O  I  N  T  A    L   I       H     O
                    E  S      A   O        Y     U
                 12 O  S  T  E  O  G  E  N  E  S  I  S
                    I          E            I
                    D                       S
```

■ Page 37

Hit or miss

1. True.
2. False. Fetal cartilage is transformed into bony skeleton by 6 months in utero.
3. True.
4. True.
5. False. Bone growth continues until the epiphyseal plates ossify, usually during late adolescence.
6. False. Bone density decreases after age 30 in women and after age 45 in men.

Strike out

2. ~~Cartilage has a rich blood supply.~~ Cartilage has no blood supply.
3. ~~Fibrous cartilage covers the articular bone surfaces and connects the ribs to the sternum.~~ Hyaline cartilage covers the articular bone surfaces and connects the ribs to the sternum.
5. ~~Joints are classified according to the size of the bone comprising the joint.~~ Joints can be classified by function (extent of movement) or by structure (what they're made of).
7. ~~Soft tissue separates the contiguous bony surfaces in synovial joints.~~ The contiguous bony surfaces in the synovial joints are separated by the synovia and by cartilage.

■ Pages 38 and 39

You make the call

1. At about the ninth month, an ossification center develops in the epiphysis. Some cartilage cells enlarge and stimulate ossification of surrounding cells. The enlarged cells die, leaving small cavities. New cartilage continues to develop.
2. Osteoblasts begin to form bone on the remaining cartilage, creating the trabeculae network of cancellous bone. Cartilage continues to form on the outer surfaces of the epiphysis and along the upper surface of the epiphyseal plate.
3. Cartilage is replaced by compact bone near the outer surface of the epiphysis. Only cartilage cells on the upper surface of the epiphyseal plate continue to multiply rapidly, pushing the epiphysis away from the diaphysis. This new cartilage ossifies, creating trabeculae on the medullary side of the epiphyseal plate.
4. Osteoclasts produce enzymes and acids that reduce trabeculae created by the epiphyseal plate, thus enlarging the medullary cavity. In the epiphysis, osteoclasts reduce bone, making its calcium available for new osteoblasts that give the epiphysis its adult shape and proportion.

■ Page 40

Train your brain

1. The continuous process whereby bone is created and destroyed is remodeling.
2. Osteoblasts deposit new bone, and osteoclasts increase long bone diameter.

■ Page 41

Match point

1. B, 2. P, 3. L, 4. F, 5. O, 6. C, 7. A, 8. J, 9. H, 10. K, 11. I, 12. D, 13. M, 14. G, 15. E, 16. N

■ Page 42

Match point

1. A, 2. D, 3. C, 4. F, 5. E, 6. B

Jumble gym

1. Synarthrosis, 2. Amphiarthrosis, 3. Diarthrosis, 4. Fibrous, 5. Cartilaginous, 6. Synovial

Answer: Articulations

Chapter 5

Page 48

Cross-training

C		²A	X	O	N	S		³N	E	U	R	O	N			
E		S						E								
R	⁴C	T			⁵H			U								
E	E	R			Y			R								
B	R	O			P			O			⁶B					
R	E	G	⁷F	R	O	N	T	A	L		A					
U	B	L			T			N			S					
M	Y	E	L	I	N			H			I					
	L	A			H			A			L			¹⁰P		
	L				A			N			A			O		
¹¹N	E	U	R	O	G	L	I	A		¹²M	I	D	B	R	A	I	N
	M				M			M			I			S		
					U			S								
¹³D	E	N	D	R	I	T	E	S		S						
								I								
			¹⁴M	I	C	R	O	G	L	I	A					
								N								

Page 49

Boxing match

1. Astroglia, 2. Ependymal, 3. Microglia, 4. Oligodendroglia

Hit or miss

1. True.
2. False. Neuron activity may be provoked by mechanical stimuli (touch and pressure), thermal stimuli (heat and cold), and chemical stimuli (external chemicals or chemicals released by the body).
3. False. Dendrites conduct impulses toward the cell body; the axon conducts impulses away from the cell body.
4. True.
5. False. Schwann cells produce the myelin sheath.
6. True.
7. True.
8. True.
9. False. The myelin sheath protects and insulates the nervous system.
10. True.

Page 50

Finish line

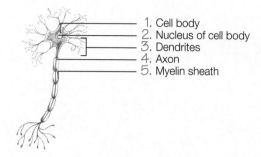

1. Cell body
2. Nucleus of cell body
3. Dendrites
4. Axon
5. Myelin sheath

Match point

1. F, 2. H, 3. J, 4. C, 5. K, 6. A, 7. E, 8. G, 9. D, 10. B, 11. I

Page 51

Power stretch

Frontal: A, D, I
Parietal: C, E, F
Temporal: B, G
Occipital: H

Finish line

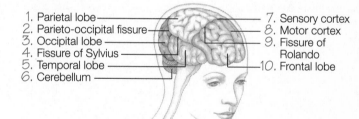

1. Parietal lobe
2. Parieto-occipital fissure
3. Occipital lobe
4. Fissure of Sylvius
5. Temporal lobe
6. Cerebellum
7. Sensory cortex
8. Motor cortex
9. Fissure of Rolando
10. Frontal lobe

■ Page 52

Circuit training

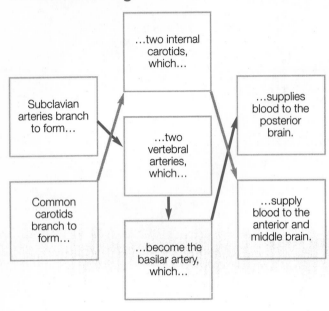

Train your brain

Blood continually circulates to the brain despite interruption of any of the brain's major vessels.

■ Page 53

Match point

1. G, 2. C, 3. J, 4. A, 5. E, 6. H, 7. B, 8. F, 9. I, 10. D

Circuit training

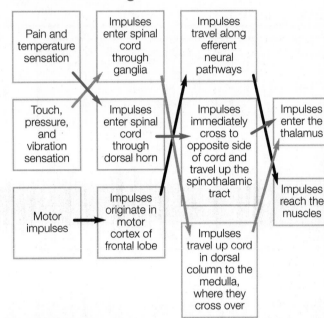

■ Page 54

Power stretch

Pyramidal: B, C, F
Extrapyramidal: A, D, E, G

Train your brain

1. Cell bodies in two dorsal horns relay sensations.
2. Cell bodies in two ventral horns are key to voluntary reflex motor activity.

■ Page 55

Parallel bars

1. Deep, 2. Extrapyramidal, 3. Afferent, 4. Thalamus, 5. Cerebellum

Match point

1. B, 2. F, 3. D, 4. G, 5. E, 6. A, 7. H, 8. I, 9. C

Page 56
You make the call

1. **Biceps reflex.** Place thumb or index finger over the biceps tendon and the remaining fingers loosely over the triceps muscle. Strike the thumb or index finger over the biceps tendon with the pointed end of the reflex hammer. Watch and feel for the contraction of the biceps muscle and flexion of the forearm.
2. **Triceps reflex.** Strike the triceps tendon about 2″ (5 cm) above the olecranon process on the extensor surface of the upper arm. Watch for contraction of the triceps muscle and extension of the forearm.
3. **Brachioradialis reflex.** Strike the radius about 1″ to 2″ (2.5 to 5 cm) above the wrist and watch for supination of the hand and flexion of the forearm at the elbow.
4. **Patellar reflex.** Strike the patellar tendon just below the patella and look for contraction of the quadriceps muscle in the thigh with extension of the leg.
5. **Achilles reflex.** With the foot flexed and supporting the plantar surface, strike the Achilles tendon. Watch for plantar flexion of the foot at the ankle.

Page 57
Hit or miss

1. True.
2. True.
3. False. There are 12 pairs of cranial nerves and 31 pairs of spinal nerves.
4. True.
5. False. The olfactory and optic nerves exit from the forebrain.
6. True.
7. False. The ANS innervates all internal organs.
8. False. The spinal nerves carry messages to particular body regions called dermatomes.

Power stretch

Sympathetic: B, D, E, G, H, K, L, N, O, P
Parasympathetic: A, C, F, I, J, M

Page 58
Finish line

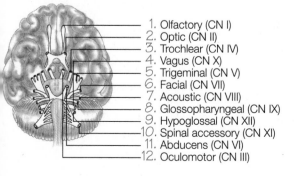

1. Olfactory (CN I)
2. Optic (CN II)
3. Trochlear (CN IV)
4. Vagus (CN X)
5. Trigeminal (CN V)
6. Facial (CN VII)
7. Acoustic (CN VIII)
8. Glossopharyngeal (CN IX)
9. Hypoglossal (CN XII)
10. Spinal accessory (CN XI)
11. Abducens (CN VI)
12. Oculomotor (CN III)

Page 59
Jumble gym

1. Sclera, 2. Cornea, 3. Iris, 4. Pupil, 5. Anterior chamber, 6. Lens, 7. Ciliary body, 8. Vitreous humor, 9. Retina, 10. Choroid
Answer: Ophthalmoscope

Page 60
Finish line

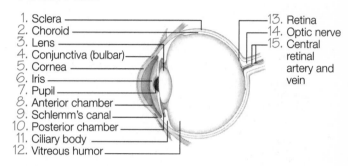

1. Sclera
2. Choroid
3. Lens
4. Conjunctiva (bulbar)
5. Cornea
6. Iris
7. Pupil
8. Anterior chamber
9. Schlemm's canal
10. Posterior chamber
11. Ciliary body
12. Vitreous humor
13. Retina
14. Optic nerve
15. Central retinal artery and vein

Match point

1. B, 2. F, 3. D, 4. H, 5. A, 6. I, 7. G, 8. J, 9. E, 10. C

■ Chapter 6

Page 64
Cross-training

(Crossword answers)
1. ACINAR, 2. AR, 3. C
4. MELATONIN
5. V
7. POLYPEPTIDES, 8. I
6. T
9. H
10. GONADS, 11. A
12. TYROSINE
13. HORMONES

Page 65

Finish line

1. Pineal gland
2. Pituitary gland
3. Thyroid gland
4. Thymus
5. Adrenal glands
6. Pancreas

Also known as...

1. Correct.
2. Incorrect. Thyroid stimulating hormone is also known as *thyrotropin*.
3. Incorrect. Luteinizing hormone and prolactin are two distinct hormones produced by the anterior pituitary gland.
4. Correct.
5. Incorrect. Triiodothyronine is T_3; T_4 is thyroxine, which together with T_3 is called *thyroid hormone*.

Page 66

Match point

1. F, 2. I, 3. L, 4. D, 5. J, 6. A, 7. G, 8. B, 9. K, 10. M, 11. E, 12. H, 13. C

Power stretch

Zona glomerulosa: A, D, E
Zona fasciculata: B, F, G, I
Zona reticularis: C, H

Page 67

Hit or miss

1. False. Acinar cells make up most of the gland and regulate pancreatic exocrine function.
2. True.
3. True.
4. False. Alpha cells produce glucagon; beta cells produce insulin.
5. True.
6. False. Insulin lowers the blood glucose level by stimulating the conversion of glucose to glycogen.
7. False. Release patterns of hormones vary greatly.
8. True.
9. True.
10. False. A particular hormone may have different effects at different target sites.

Power stretch

Polypeptides: A, C, F, H
Steroids: D, G
Amines: B, E

Page 68

You make the call

1. Simple feedback occurs when the level of one substance regulates the secretion of hormones (simple loop). For example, a low serum calcium level stimulates the parathyroid gland to release parathyroid hormone (PTH). PTH, in turn, promotes resorption of calcium from the GI tract, kidneys, and bones. A high serum calcium level inhibits PTH secretion.
2. When the hypothalamus receives negative feedback from target glands, the mechanism is more complicated (complex loop). Complex feedback occurs through an axis established between the hypothalamus, pituitary gland, and target organ. For example, secretion of corticotropin-releasing hormone from the hypothalamus stimulates release of corticotropin by the pituitary gland, which in turn stimulates cortisol secretion by the adrenal gland (the target organ). A rise in serum cortisol levels inhibits corticotropin secretion by decreasing corticotropin-releasing hormone.

Strike out

3. ~~By increasing tropic hormones, the pituitary gland decreases target gland hormone levels.~~ Increased trophic hormones stimulate the target gland to increase production of target hormones; decreased trophic hormones decrease target gland stimulation and target gland hormone levels.
5. ~~The ability of blood glucose levels to regulate glucagon and insulin release is an example of the hypothalamic-pituitary-target gland axis.~~ The ability of blood glucose levels to regulate glucagon and insulin release is an example of chemical regulation.

Chapter 7

Page 72

Finish line

1. Superior vena cava
2. Branches of right pulmonary artery
3. Right atrium
4. Right pulmonary veins
5. Tricuspid valve
6. Chordae tendineae
7. Right ventricle
8. Papillary muscle
9. Inferior vena cava
10. Pulmonary semilunar valve
11. Aortic arch
12. Branches of left pulmonary artery
13. Left atrium
14. Left pulmonary veins
15. Mitral valve
16. Aortic semilunar valve
17. Myocardium
18. Left ventricle
19. Interventricular septum
20. Descending aorta

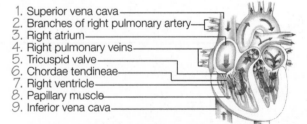
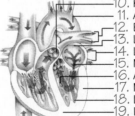

■ Page 73

Match point

1. B, 2. E, 3. H, 4. F, 5. A, 6. D, 7. G, 8. C, 9. I

Circuit training

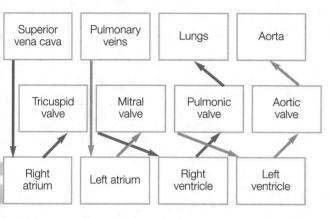

■ Page 74

Cross-training

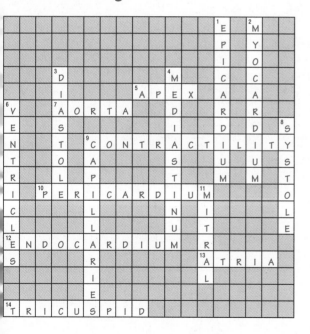

■ Page 75

Strike out

2. ~~The AV valves separate the aorta from the ventricles.~~ The AV valves separate the atria from the ventricles.
3. ~~The right AV valve is called the *mitral valve*.~~ The right AV valve is called the *tricuspid valve;* the left AV valve is called the *mitral valve.*
5. ~~The left AV valve prevents backflow into the left ventricle from the aorta.~~ The left AV valve prevents backflow from the left ventricle into the left atrium.
7. ~~The pulmonic valve prevents backflow from the pulmonary veins into the right atrium.~~ The pulmonic valve prevents backflow from the pulmonary artery into the right ventricle.
10. ~~Both the tricuspid and mitral valve have three cusps, or leaflets.~~ The tricuspid valve has three cusps; the mitral valve has two.

Train your brain

The heart valves open and close in response to pressure changes caused by ventricular contraction and blood ejection.

■ Page 76

Parallel bars

1. Wall, 2. Parietal and visceral, 3. Endo, 4. Atrium, 5. Mitral (or bicuspid), 6. Automaticity, 7. 40 to 60

Hit or miss

1. True.
2. False. The SA node is located on the endocardial surface of the right atrium, near the superior vena cava.
3. False. The SA node generates an impulse between 60 and 100 times per minute.
4. True.
5. False. The AV node is located low in the septal wall of the right atrium.
6. False. The AV node slows impulse conduction to allow time for the contracting atria to fill the ventricles with blood before the lower chambers contract.
7. True.
8. True.
9. True.
10. False. If the SA node fails to fire, the AV node will generate an impulse between 40 and 60 times per minute.
11. False, If the AV node fails to fire, the ventricles can generate their own impulse between 20 and 40 times per minute.

■ Page 77

Finish line

1. Interatrial tract
2. Sinoatrial node
3. Internodal tracts
4. Atrioventricular node
5. Bundle of His (atrioventricular bundle)
6. Right bundle branch
7. Left bundle branch
8. Purkinje fibers

Page 78

Train your brain

The cardiac cycle is the period of time from the beginning of one heartbeat to the beginning of the next.

Match point

1. C, 2. F, 3. H, 4. A, 5. D, 6. G, 7. I, 8. B, 9. E

Page 79

You make the call

1. In response to ventricular depolarization, tension in the ventricles increases. This rise in pressure within the ventricles leads to closure of the mitral and tricuspid valves. The pulmonic and aortic valves stay closed during the entire phase.
2. When ventricular pressure exceeds aortic and pulmonary arterial pressure, the aortic and pulmonic valves open and the ventricles eject blood.
3. When ventricular pressure falls below the pressure in the aorta and pulmonary artery, the aortic and pulmonic valves close. All valves are closed during this phase. Atrial diastole occurs as blood fills the atria.
4. Atrial pressure exceeds ventricular pressure, which causes the mitral and tricuspid valves to open. Blood then flows passively into the ventricles. About 70% of ventricular filling takes place during this phase.
5. Atrial systole (coinciding with late ventricular diastole) supplies the ventricles with the remaining 30% of the blood for each heartbeat.

Page 80

Power stretch

Preload: C, F
Contractility: A, E
Afterload: B, D

Odd man out

1. Lungs. The heart, blood vessels, and lymphatics are all components of the cardiovascular system. The lungs aren't.
2. Pericardium. The epicardium, endocardium, and myocardium are layers of the heart wall. The pericardium is the sac that surrounds the heart.
3. Mediastinum. The atria and ventricles are the chambers of the heart. The mediastinum is the cavity between the lungs in which the heart is located.
4. Pulmonic. The bicuspid valve (also known as the *mitral valve*) and the tricuspid valve are AV valves; the pulmonic valve is a semilunar valve.
5. Alveoli. Arteries, venules, veins, arterioles, and capillaries are types of blood vessels. Alveoli are sacs in the terminal bronchioles of the lungs, in which carbon dioxide and oxygen exchange occurs.
6. Intrinsic. Systemic, coronary, and pulmonary are the three methods of circulation that carry blood throughout the body.

Page 81

Jumble gym

1. Arteries, 2. Arterioles, 3. Venules, 4. Veins, 5. Sphincters, 6. Valves
Answer: Capillaries

Circuit training

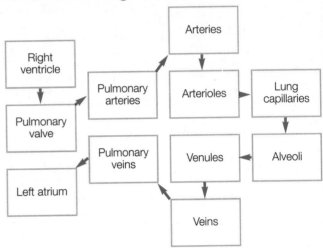

Page 82

Finish line

1. Transverse sinus
2. Right jugular vein
3. Brachiocephalic vein
4. Superior vena cava
5. Right atrium
6. Right ventricle
7. Inferior vena cava
8. Renal veins
9. Common iliac vein
10. Femoral vein
11. Popliteal vein

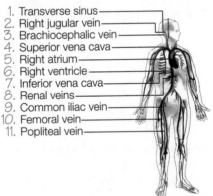

12. Right common carotid artery
13. Temporal artery
14. Left common carotid artery
15. Left subclavian artery
16. Left atrium
17. Left ventricle
18. Aorta
19. Renal arteries
20. Common iliac artery
21. Radial artery
22. Ulnar artery
23. Femoral artery
24. Popliteal artery
25. Posterior tibial artery
26. Dorsalis pedis artery

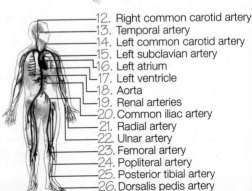

■ Page 83

Train your brain

Coronary arteries and their branches supply oxygenated blood to the heart while cardiac veins remove oxygen-depleted blood.

Power stretch

Right coronary artery: B, D, F, I
Left coronary artery: A, G, K, L
Cardiac veins: C, H
Coronary sinus: E, J

■ Page 84

Finish line

Anterior view

1. Left coronary artery
2. Right coronary artery
3. Anterior cardiac veins
4. Small cardiac vein
5. Circumflex branch of left coronary artery
6. Great cardiac vein
7. Anterior descending branch of left coronary artery

Posterior view

1. Great cardiac vein
2. Coronary sinus
3. Posterior vein of left ventricle
4. Middle cardiac vein
5. Posterior descending branch of right coronary artery

■■
■ Chapter 8

■ Page 88

Cross-training

													¹V		²P
												³R		A		L
		⁴H	E	M	A	T	O	P	O	I	E	S	I	S		A
		E						T			C		O		T
⁵L	Y	M	P	H	O	C	Y	T	E		I		C		E
		O						I			C		O		L
⁶H	I	S	T	A	M	I	N	E			U		N		E
		T						L			S		S		T
		A		⁷E	R	Y	T	H	R	O	C	Y	T	E	S
		S						C			R		I		
		I						Y			I				
		S		⁸N				T			C				
			⁹L	E	U	K	O	C	Y	T	E	S	T		¹⁰M
			U									I		O
			T			¹¹H	E	M	O	G	L	O	B	I	N
			R								N		O
¹²P	H	A	G	O	C	Y	T	E					C
			P										Y
			H										T
	¹³A	N	T	I	G	E	N	S					E
			L										

■ Page 89

Parallel bars

1. Thrombocyte. *Erythrocyte* is another name for a red blood cell; *thrombocyte* is another name for a platelet.
2. Agranulocyte. An eosinophil is a type of granulocyte, and a lymphocyte is a type of agranulocyte.
3. Band. A reticulocyte is an immature red blood cell; a band is an immature neutrophil, a type of white blood cell.
4. Fibrinogen. Factor I is also called *fibrinogen;* factor XIII is also called *fibrin stabilizing factor.*

Strike out

3. ~~The type of hemoglobin in the RBC determines a person's blood type.~~ Antigens on the RBC surface determine a person's blood group, or blood type.
4. ~~RBCs have an average life span of 12 months.~~ RBCs have an average life span of 120 days.
7. ~~The rate of reticulocyte release is fixed from birth.~~ The rate of reticulocyte release usually equals the rate of old RBC removal.

■ **Page 90**

Power stretch

Neutrophils: B, D, F, H, K
Eosinophils: A, E, I, L
Basophils: C, G, J

■ **Page 91**

Hit or miss

1. True.
2. False. Testing for the presence of A and B antigens on RBCs is the most important system for classifying blood.
3. False. Type A blood has A antigen but no A antibodies; however, it does have B antibodies.
4. True.
5. False. Blood typically contains the Rh antigen, and is known as Rh-positive. Blood without the Rh antigen is Rh-negative.
6. False. Type O blood is compatible with type O only.

Match point

1. C, 2. D, 3. H, 4. A, 5. L, 6. I, 7. M, 8. F, 9. B, 10. N, 11. J, 12. E, 13. K, 14. G

■ **Page 92**

Interval training

Blood group	Antibodies present in plasma	Compatible RBCs	Compatible plasma
Recipient			
O	Anti-A and anti-B 1	O 8	O, A, B, AB
A	Anti-B 2	A, O 9	A, AB 15
B	Anti-A 3	B, O 10	B, AB 16
AB	Neither anti-A nor anti-B	AB, A, B, O	AB 17
Donor			
O	Anti-A and anti-B 4	O, A, B, AB 11	O 18
A	Anti-B 5	A, AB 12	A, O 19
B	Anti-A 6	B, AB 13	B, O 20
AB	Neither anti-A nor anti-B 7	AB 14	AB, A, B, O 21

Chapter 9

■ **Page 97**

Finish line

1. Tonsils
2. Thymus
3. Lymphatic vessels and blood capillaries
4. Spleen
5. Peyer's patches
6. Appendix
7. Bone marrow

Train your brain

The bone marrow and the thymus each play a role in the development of B cells and T cells.

■ **Page 98**

Cross-training

Crossword solution:

- 1. L (LYMPHNODE / LYMMU...)
- 2. M (MULTIPOTENT)
- 3. T (THYMUS)
- 4. HYPERSENSITIVITY
- 5. I
- 6. N (NEU...)
- 7. ANTIBODY / ANTIGEN
- 8. A (AUFCY...)
- 9. ANTITOXIN
- 10. HEMATOPOIESIS
- 11. IMMUNITY
- 12. SPLEEN
- 13. PHAGOCYTES

■ Page 99

Strike out

3. ~~Those that become phagocytes are further differentiated to become either B cells or T cells.~~ Those that become lymphocytes are further differentiated to become either B cells or T cells.

5. ~~Once mature, B cells and T cells remain in either the bone marrow or the thymus.~~ B cells and T cells are distributed throughout the lymphoid organs, especially the lymph nodes and spleen.

7. ~~The thymus remains active throughout adulthood.~~ The thymus only has function in the body's immunity for the first several months after birth.

Memory jogger

1. **B**one marrow
2. **T**hymus

■ Page 100

Match point

1. C, 2. F, 3. A, 4. G, 5. I, 6. D, 7. J, 8. H, 9. B, 10. E

Finish line

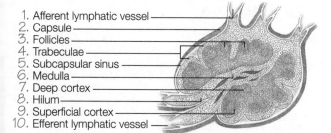

1. Afferent lymphatic vessel
2. Capsule
3. Follicles
4. Trabeculae
5. Subcapsular sinus
6. Medulla
7. Deep cortex
8. Hilum
9. Superficial cortex
10. Efferent lymphatic vessel

■ Page 101

A-maze-ing race

If you answered all of the questions correctly, Mr. Spleen should have ended up at door number 30. The answers to the individual questions are:

1. True.
2. False.
3. False.
4. True.
5. True.
6. True.
7. False.

■ Page 102

Hit or miss

1. False. Intact and healing skin and mucous membranes provide the first line of defense against microbial invasion.
2. False. Skin desquamation and low pH help impede bacterial colonization.
3. True.
4. True.
5. True.
6. False. Nasal secretions contain an immunoglobulin that discourages microbe adherence.
7. True.
8. False. The low pH of gastric secretions renders the stomach virtually free from live bacteria. Also, in the rest of the GI system, resident bacteria prevent other microorganisms from developing colonies.
9. True.

Circuit training

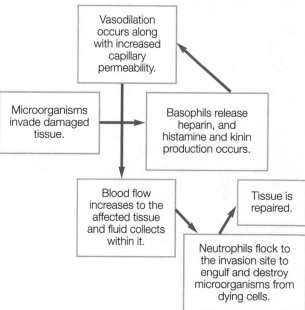

■ Page 103

Power stretch

Neutrophils: A, C, G, J, K
Eosinophils: D, H
Basophils: B, F, I
Mast cells: E, F, I

■ Page 104

Match point

1. C, 2. E, 3. G, 4. A, 5. I, 6. F, 7. J, 8. B, 9. K, 10. H, 11. D

■ Page 105

You make the call

Chemotaxis
Chemotactic factors attract macrophages to the antigen site.

Opsonization
Antibody (immunoglobulin G) or complement fragment coats the microorganism, enhancing macrophage binding to the antigen, now called an *opsinogen*.

Ingestion
The macrophage extends its membrane around the opsonized microorganism, engulfing it within a vacuole (phagosome).

Digestion
As the phagosome shifts away from the cell periphery, it merges with lysosomes, forming a phagolysosome, where antigen destruction occurs.

Release
When digestion is complete, the macrophage expels digestive debris, including lysosomes, prostaglandins, complement components, and interferon, which continue to mediate the immune response.

■ Page 106

Strike out

1. ~~T cells are involved in humoral immunity.~~ T cells are involved in cell-mediated immunity; B cells are involved in humoral immunity.
3. ~~IgE acts as an antigen receptor of B cells.~~ IgE causes an allergic reaction; IgD acts as an antigen receptor of B cells.
4. ~~All immunoglobulins work to disable bacteria by producing antitoxins.~~ Immunoglobulins can work in one of several ways, depending on the antigen. This includes producing antitoxins, coating the bacteria to make them targets for phagocytosis, or linking to the antigen to trigger the production of complement.
6. ~~During a secondary antibody response, memory B cells again manufacture IgM antibodies, this time achieving peak levels in 1 to 2 days.~~ During a secondary antibody response, memory B cells manufacture mainly IgG antibodies, which achieve peak levels in 1 to 2 days.
9. ~~One of the by-products of the complement cascade is to produce more T cells.~~ By-products of the complement cascade produce the inflammatory response, stimulation and attraction of neutrophils, and coating of target cells to make them attractive to phagocytes.

Match point

1. B, 2. D, 3. E, 4. A, 5. F, 6. C

■■ ■ Chapter 10

■ Page 112

Batter's box

1. upper, 2. lower, 3. lungs, 4. thoracic, 5. nose, 6. mouth, 7. larynx, 8. trachea, 9. bronchi, 10. lungs, 11. oxygen, 12. carbon dioxide, 13. acid-base

■ Page 113

Cross-training

	¹C	O	N	²C	H	A	E							
				Y						³L	⁴A	R	Y	N	X
				P							C			
⁵H	Y	P	O	V	E	N	T	I	L	A	T	I	O	N
I				R					N			⁶P		
L				V					U			L		
U				E		⁷V	I	B	R	I	S	S	A	E
M				N					U			U		
		⁹S	E	P	T	U	M					R	R	
				I								F	A	
				L								A		
		¹⁰C	A	R	I	N	A					C		
				T				¹¹N				T		
				I				A				A		
		¹²O	R	O	P	H	A	R	Y	N	X			
				N				E				T		
				S				S						

■ Page 114

Finish line

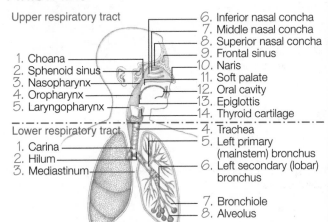

Upper respiratory tract
1. Choana
2. Sphenoid sinus
3. Nasopharynx
4. Oropharynx
5. Laryngopharynx
6. Inferior nasal concha
7. Middle nasal concha
8. Superior nasal concha
9. Frontal sinus
10. Naris
11. Soft palate
12. Oral cavity
13. Epiglottis
14. Thyroid cartilage

Lower respiratory tract
1. Carina
2. Hilum
3. Mediastinum
4. Trachea
5. Left primary (mainstem) bronchus
6. Left secondary (lobar) bronchus
7. Bronchiole
8. Alveolus

■ Page 115

Train your brain

The four paranasal sinuses provide speech resonance.

Jumble gym

1. Cartilage, 2. Cilia, 3. Conchae, 4. Conducting airways, 5. Cricoid cartilage
Answer: Choanae

■ Page 116

Hit or miss

1. True.
2. True.
3. False. The primary bronchi begin at the carina. The secondary bronchi enter the pleural cavities and the lungs at the hilum
4. True.
5. False. The right lung has three lobes; the left has two lobes.
6. False. The lobar bronchi branch into segmental bronchi (tertiary bronchi). In turn, these segments branch into smaller and smaller bronchi, finally branching into bronchioles.
7. False. The larger bronchi consist of cartilage, smooth muscle, and epithelium.
8. True.
9. True.
10. True.
11. False. Type I cells are the most abundant, but they don't secrete surfactant. It's across these cells that gas exchange takes place. Type II cells secrete surfactant.
12. True.
13. True.

Finish line

1. Terminal bronchioles
2. Respiratory bronchioles
3. Alveolus
4. Alveolar sac
5. Acinus

■ Page 117

Match point

1. C, 2. D, 3. G, 4. H, 5. F, 6. B, 7. I, 8. E, 9. A

Strike out

1. ~~The left lung is larger than the right and handles the majority of gas exchange.~~ The right lung is shorter, broader, and larger than the left. It handles 55% of gas exchange.
3. ~~Another function of serous fluid is to prevent inflammation on the lung surface.~~ Serous fluid has no anti-inflammatory role. Rather, its other function is to create a bond between the visceral and parietal layers. This bond causes the lungs to move with the chest wall during breathing.
5. ~~Cartilage forms the major portion of the thoracic cage.~~ The ribs form the major portion of the thoracic cage.
8. ~~All of the ribs attach directly to the sternum.~~ Ribs 1 through 7 attach directly to the sternum; ribs 8 through 10 attach to the cartilage of the preceding rib; the other 2 pairs of ribs are "free-floating."
10. ~~The costal angle is the angle of about 90 degrees that forms at the point where the bottom rib meets the 12th thoracic vertebra.~~ The costal angle is the angle of about 90 degrees that forms near the xiphoid process at the lower part of the rib cage.

■ Page 118

Finish line

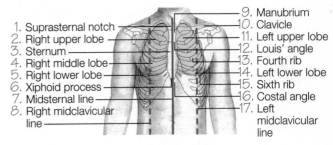

1. Suprasternal notch
2. Right upper lobe
3. Sternum
4. Right middle lobe
5. Right lower lobe
6. Xiphoid process
7. Midsternal line
8. Right midclavicular line
9. Manubrium
10. Clavicle
11. Left upper lobe
12. Louis' angle
13. Fourth rib
14. Left lower lobe
15. Sixth rib
16. Costal angle
17. Left midclavicular line

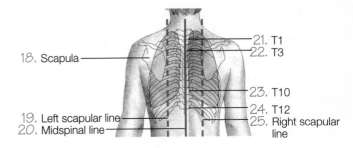

18. Scapula
19. Left scapular line
20. Midspinal line
21. T1
22. T3
23. T10
24. T12
25. Right scapular line

■ Page 119

You make the call

1. Before inspiration, intrapulmonary pressure equals atmospheric pressure (approximately 760 mm Hg). Intrapleural pressure is 756 mm Hg.
2. During inspiration, the diaphragm and external intercostal muscles contract, enlarging the thorax vertically and horizontally. As the thorax expands, intrapleural pressure decreases and the lungs expand to fill the enlarging thoracic cavity.
3. The intrapulmonary atmospheric pressure gradient pulls air into the lungs until the two pressures are equal.
4. During normal expiration, the diaphragm slowly relaxes and the lungs and thorax passively return to resting size and position. During deep or forced expiration, contraction of internal intercostal and abdominal muscles reduces thoracic volume. Lung and thorax compression raises intrapulmonary pressure above atmospheric pressure.

■ Page 120

Mind sprints

1. With age, the nose tends to enlarge from continued cartilage growth.
2. Anteroposterior chest diameter increases as a result of altered calcium metabolism.
3. Calcification of costal cartilages results in decreased mobility of the chest wall.
4. The lungs become more rigid and the number and size of alveoli decline.
5. Respiratory fluids decline by 30%, increasing the risk of pulmonary infection and mucus plugs.
6. The lung's diffusing capacity declines while decreased inspiratory and expiratory muscle strength diminishes vital capacity.
7. Lung tissue degeneration causes a decrease in the lungs' elastic recoil capability, which results in an elevated residual volume.
8. Some airways close, leading to poor ventilation of the basal areas and a decreased partial pressure of oxygen in arterial blood.

Power stretch

1. Forced inspiration: A, C, D, F
2. Active expiration: B, E

■ Page 121

Train your brain

Effective respiration consists of gas exchange in the lungs, called *external respiration,* and gas exchange in the tissues, called *internal respiration.*

■ Page 122

Circuit training

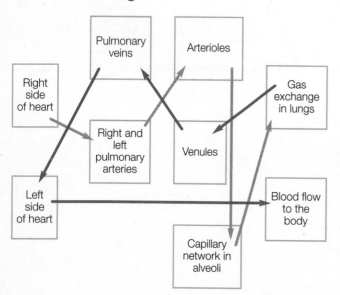

■ Page 123

Strike out

3. ~~The pons controls the rate and depth of respiration.~~ The medulla controls the rate and depth of respiration. The pons moderates the rhythm of the switch from inspiration to expiration.
4. ~~The pons stimulates contraction of the diaphragm and external intercostal muscles, which produce the intrapulmonary pressure changes that cause inspiration.~~ The medulla stimulates contraction of the diaphragm and external intercostal muscles.
7. ~~Perfusion aids internal respiration.~~ Perfusion aids external respiration.
8. ~~Gravity interferes with perfusion by causing blood to pool in the bases of the lungs.~~ Gravity enhances oxygen and carbon dioxide transportation by causing more unoxygenated blood to travel to the lower and middle lung lobes.
9. ~~In a healthy lung, ventilation and perfusion should be the same throughout the lung.~~ Ventilation and perfusion normally differ in various parts of the lungs.

Mind sprints

1. Heart and pericardium
2. Thoracic aorta
3. Pulmonary arteries and veins
4. Venae cavae and azygos veins
5. Thymus
6. Lymph nodes
7. Trachea
8. Esophagus
9. Thoracic duct
10. Vagus, cardiac, and phrenic nerves

■ Page 124

You make the call

1. This illustrates laminar flow. The linear pattern of laminar flow occurs at low flow rates and offers minimal resistance. This flow type occurs mainly in the small peripheral airways of the bronchial tree.
2. This illustrates turbulent flow. The eddying pattern of turbulent flow creates friction and increases resistance. Turbulent flow is normal in the trachea and large central bronchi. If the smaller airways become constricted or clogged with secretions, however, turbulent flow may also occur there.
3. This illustrates transitional flow, which is a mixed pattern common at lower flow rates in the larger airways, especially where the airways narrow from obstruction, meet, or branch.

■ Page 125

Match point

1. B, 2. F, 3. D, 4. A, 5. H, 6. E, 7. I, 8. C, 9. G

Hit or miss

1. True.
2. False. Perfusion aids external respiration.
3. False. Both the alveolar epithelium and the capillary endothelium are composed of a single layer of cells. Between these layers are tiny interstitial spaces filled with elastin and collagen.
4. True.
5. False. When oxygen arrives in the bloodstream, it displaces carbon dioxide.
6. False. Most transported oxygen binds with hemoglobin to form oxyhemoglobin. However, a small portion dissolves in the plasma. (This can be measured as the partial pressure of oxygen in arterial blood, or Pao_2.)
7. True.
8. True.
9. True.

■ Page 126

You make the call

1. This shows what occurs in a normal lung, in which ventilation closely matches perfusion
2. This illustrates a shunt, the result of perfusion without adequate ventilation. A shunt usually results from airway obstruction, particularly that caused by acute diseases, such as atelectasis and pneumonia.
3. This illustrates dead-space ventilation, a result of normal ventilation without adequate perfusion. This is usually caused by a perfusion defect such as pulmonary embolism.
4. This shows both inadequate ventilation and perfusion (silent unit). This usually stems from multiple causes such as pulmonary embolism with resultant adult respiratory distress syndrome and emphysema.

■ Chapter 11

■ Page 132

Cross-training

		¹R													
²I	L	E	U	M									³C		
		C											H		
		T								⁴J		Y			
	⁵D	U	O	D	E	N	U	⁶M		⁷C	E	C	U	M	
		M						A		J			E		
			⁸H			S		U							
	⁹D	E	G	L	U	T	I	T	I	O	N				
		P				I		U							
		A				C		M							
		T		¹⁰P	H	A	R	Y	N	X		¹¹M			
¹²L		O				T				U					
¹³S	T	O	M	A	C	H		¹⁴V	I	L	L	I		C	
B		Y				O				O					
U		T		¹⁵S	I	N	U	S	O	I	D	S			
L		E								S		A			
E	¹⁶E	S	O	P	H	A	G	U	S						

■ Page 133

Finish line

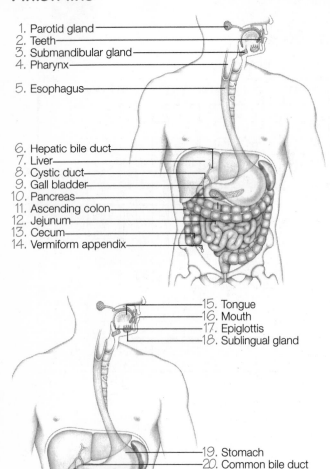

1. Parotid gland
2. Teeth
3. Submandibular gland
4. Pharynx
5. Esophagus

6. Hepatic bile duct
7. Liver
8. Cystic duct
9. Gall bladder
10. Pancreas
11. Ascending colon
12. Jejunum
13. Cecum
14. Vermiform appendix

15. Tongue
16. Mouth
17. Epiglottis
18. Sublingual gland

19. Stomach
20. Common bile duct
21. Duodenum
22. Duodenojejunal flexure
23. Transverse colon
24. Descending colon
25. Ileum
26. Signoid colon
27. Rectum

■ Page 134

Match point

1. C, 2. B, 3. D, 4. A, 5. E

■ Page 135

Power stretch

1. Stomach: B, D, G, I
2. Small intestine: C, F, J
3. Large intestine: A, E H

Train your brain

The mouth initiates the mechanical breakdown of food.

■ Page 136

Match point

1. B, 2. C, 3. A, 4. D

Train your brain

1. G cells in the pyloric glands secrete gastrin.
2. S cells in the duodenal and jejunal glands secrete secretin.

■ Page 137

Finish line

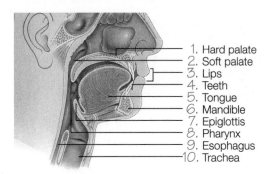

1. Hard palate
2. Soft palate
3. Lips
4. Teeth
5. Tongue
6. Mandible
7. Epiglottis
8. Pharynx
9. Esophagus
10. Trachea

■ Page 138

Power stretch

Mucosa: A, D, E, G
Submucosa: C, F, H
Tunica muscularis: B, I

Match point

1. D, 2. C, 3. E, 4. A, 5. F, 6. B

■ Page 139

Strike out

3. ~~Parasympathetic stimulation of the GI tract relaxes the smooth muscle of the GI tract.~~ Parasympathetic stimulation of the vagus and sacral spinal nerves increases gut and sphincter tone.
5. ~~The visceral peritoneum lines the abdominal cavity.~~ The parietal peritoneum lines the abdominal cavity.

Hit or miss

1. False. The small intestine is the longest organ of the GI tract.
2. True.
3. False. The lesser omentum is a fold of the peritoneum that covers most of the liver and anchors it to the lesser curvature of the stomach.
4. False. The liver consists of four lobes: left lobe, right lobe, caudate lobe, and quadrate lobe.
5. False. The liver's functional unit is called a *lobule*.
6. True.

■ Page 140

Mind sprints

1. Plays an important role in carbohydrate metabolism
2. Detoxifies various endogenous and exogenous toxins in plasma
3. Stores essential nutrients, such as vitamins K, D, and B_{12}, and iron
4. Synthesizes plasma proteins, nonessential amino acids, and vitamin A
5. Removes ammonia from body fluids, converting it to urea for excretion in urine
6. Helps regulate blood glucose levels
7. Secretes bile

Match point

1. C, 2. D, 3. A, 4. E, 5. B

■ Page 141

Finish line

1. Right and left hepatic ducts
2. Cystic duct
3. Neck of gallbladder
4. Body of gallbladder
5. Fundus of gallbladder
6. Minor duodenal papilla
7. Sphincter muscles
8. Major duodenal papilla
9. Duodenum

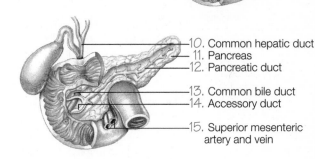

10. Common hepatic duct
11. Pancreas
12. Pancreatic duct
13. Common bile duct
14. Accessory duct
15. Superior mesenteric artery and vein

Match point

1. C, 2. E, 3. G, 4. B, 5. H, 6. A, 7. F, 8. D, 9. I,

■ Page 142

You make the call

1. Food pushed to the back of the mouth simulates swallowing receptor areas that surround the pharyngeal opening.
2. These receptor areas transmit impulses to the brain by way of the sensory portions of the trigeminal and glossopharyngeal nerves.
3. The brain's swallowing center then relays motor impulses to the esophagus by way of the trigeminal, glossopharyngeal, vagus, and hypoglossal nerves, causing swallowing to occur.

Hit or miss

1. True.
2. False. The projections of the intestinal mucosa increase the surface area for absorption.
3. True.
4. True.
5. False. The intestinal phase of digestion begins when chyme is released into the duodenum.

■ Page 143

Starting lineup

> 1. Chewing, salivation, and swallowing occur.

> 2. Food bolus enters the esophagus and is lubricated with mucus.

> 3. Glossopharyngeal nerve activates peristalsis, moving the food toward the stomach.

> 4. The stomach stretches.

> 5. Gastrin is released.

> 6. Gastric juices mix with and break down the food, forming chyme.

> 7. Chyme moves into the antrum and then on to the duodenum.

> 8. Intestinal contractions and various digestive secretions break down carbohydrates, proteins, and fats.

> 9. Bolus passes through the large intestine and on to the sigmoid colon.

> 10. Excretion exits through the anal canal.

Mind sprints

1. Gastrin release
2. Stomach distention
3. Enterogastric reflex

■ Page 144

Hit or miss

1. True.
2. False. The internal sphincter is under autonomic control; the external sphincter is under voluntary control.
3. False. The large intestine produces no hormones or enzymes; it continues the absorptive process.
4. True.
5. False. The large intestine harbors bacteria that help synthesize vitamin K and break down cellulose into a usable carbohydrate.

Train your brain

When bile salts are absent from the intestinal tract, lipids are excreted and fat-soluble vitamins are poorly absorbed.

■ Chapter 12

■ Page 150

Batter's box

1. nutrition, 2. components, 3. chemical reactions, 4. chemical reactions, 5. energy, 6. anabolism, 7. catabolism, 8. water, 9. nutrients, 10-12. lipids, carbohydrates, proteins, 13-14. vitamins, minerals

■ Page 151

Cross-training

¹V		²L									
I	³M	A	L	T	O	S	E				
T		C									
A		T					⁴P				
⁵M	O	N	O	S	A	C	C	⁶H	A	R	I	D	E	⁷S
I		S					Y	O			U
N		E					D	T			C
S							R	E			R
							O	I			O
		⁸I	N	S	U	L	I	N			S
							Y	S			E
	⁹A	M	Y	L	A	S	E			
						I			
	¹⁰G	L	U	C	O	S	E		

■ Page 152

Power stretch

Proteins: B, E, F, I
Lipids: A, C, G, K
Carbohydrates: D, F, H, J

Memory jogger

1. **P**eptide
2. **P**olypeptide
3. **P**rotein

■ Page 153

Match point

1. C, 2. B., 3. G, 4. D, 5. I, 6. F, 7. J, 8. A, 9. E, 10. K, 11. H

■ Page 154

Hit or miss

1. True.
2. True.
3. True
4. False. The liver converts amino acids not needed for protein synthesis into peptides.
5. False. Most fat digestion occurs in the small intestine.
6. True.
7. False. Most of the carbohydrates in absorbed food is quickly catabolized for the release of energy.

Train your brain

All ingested carbohydrates are converted to glucose, the body's main energy source.

■ Page 155

You make the call

1. **Glycolysis.** Glycolysis breaks apart one molecule of glucose to form two molecules of pyruvate, which yields energy in the form of adenosine triphosphate and acetyl coenzyme A (CoA).
2. **Krebs cycle.** In the Krebs cycle, fragments of acetyl CoA join to oxaloacetic acid to form citric acid. The CoA molecule breaks off from the acetyl group and may form more acetyl CoA molecules. Citric acid is first converted into intermediate compounds and then back into oxaloacetic acid. The Krebs cycle also liberates carbon dioxide.

■ Page 156

Strike out

3. ~~The body stores excess amino acids in the liver.~~ The body can't store amino acids.
4. ~~Amino acids are classified as essential or nonessential based on whether they are needed for protein synthesis.~~ Amino acids are classified as essential or nonessential based on whether the human body can synthesize them.
6. ~~Use of fat instead of glucose for energy leads to an amino acid deficit.~~ Use of fat instead of glucose for energy leads to an excess of ketone bodies.

Train your brain

Lipids are stored in adipose tissue within cells until needed for use as fuel.

■ ■ Chapter 13

■ Page 160

Cross-training

■ Page 161

Train your brain

The kidneys are bean-shaped, highly vascular organs.

Strike out

2. ~~The kidneys regulate the chemical composition of blood by producing sodium, potassium, and phosphorus.~~ The kidneys regulate the chemical composition of the blood by selectively reabsorbing and secreting ions, depending upon the body's needs.

5. ~~The kidneys help maintain glucose levels by stimulating the release of insulin from the pancreas.~~ The kidneys have no role in the regulation of blood glucose levels.

7. ~~The kidneys convert vitamin A to a more active form.~~ The kidneys convert vitamin D to a more active form.

■ Page 162

Finish line

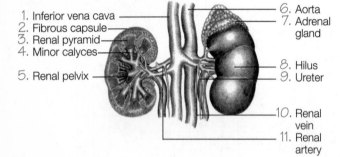

1. Inferior vena cava
2. Fibrous capsule
3. Renal pyramid
4. Minor calyces
5. Renal pelvis
6. Aorta
7. Adrenal gland
8. Hilus
9. Ureter
10. Renal vein
11. Renal artery

Hit or miss

1. False. The right kidney is situated slightly lower than the left to make room for the liver.
2. False. The position of the kidneys shifts somewhat with changes in body position.
3. True.
4. True.
5. False. Filtered blood returns to circulation by way of the renal vein, which empties into the inferior vena cava.
6. True.
7. True.

■ Page 163

Power stretch

Renal cortex: B, D, E
Renal medulla: C, F
Renal pelvis: A, G

Match point

1. D, 2. F, 3. A, 4. C, 5. E, 6. B

■ Page 164

Circuit training

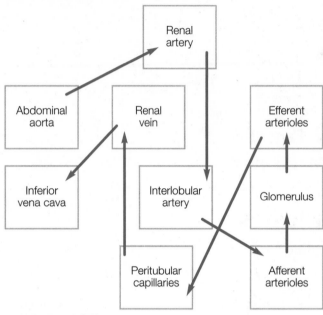

■ Page 165

Finish line

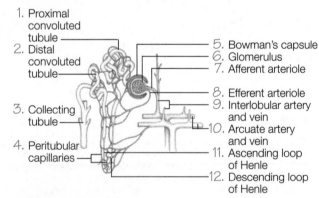

1. Proximal convoluted tubule
2. Distal convoluted tubule
3. Collecting tubule
4. Peritubular capillaries
5. Bowman's capsule
6. Glomerulus
7. Afferent arteriole
8. Efferent arteriole
9. Interlobular artery and vein
10. Arcuate artery and vein
11. Ascending loop of Henle
12. Descending loop of Henle

■ Page 166

Jumble gym

1. Glomerulus, 2. Proximal convoluted tubule, 3. Loop of Henle, 4. Distal convoluted tubule
Answer: Renal cortex

Train your brain

The loops of Henle, together with their blood vessels and collecting tubules, form the renal pyramids.

Page 167

Hit or miss

1. True.
2. True.
3. False. By the time the filtrate enters the descending limb of the loop of Henle, its water content has been reduced by 70%.
4. False. At this point the filtrate contains a high concentration of salts, chiefly sodium.
5. True.
6. True.
7. True.

Page 168

Match point

1. C, 2. D, 3. B, 4. E, 5. A

Batter's box

1. 500, 2. 600, 3. symphysis pubis, 4. ureters, 5. urethra, 6. involuntary, 7. voluntary, 8. parasympathetic, 9. internal, 10. cerebrum, 11. external, 12. micturition

Page 169

Jumble gym

1. Antidiuretic hormone, 2. Angiotensin I, 3. Angiotensin II, 4. Aldosterone, 5. Erythropoietin

Answer: Renin

Hit or miss

1. False. High levels of ADH increase water absorption and urine concentration, whereas lower levels of ADH decrease water absorption and dilute urine.
2. True.
3. False. Angiotensin I is converted to angiotensin II. Angiotensin II has a constricting effect on the arterioles, thus raising blood pressure.
4. True.
5. True.
6. False. The adrenal cortex increases aldosterone secretion when potassium levels rise. This, in turn, causes sodium retention, thereby raising blood pressure.
7. True.

Page 170

You make the call

Step 1: As blood flows into the glomerulus, filtration occurs. Active transport from the proximal convoluted tubules leads to reabsorption of sodium and glucose into nearby circulation. Osmosis then causes water reabsorption.
Step 2: In tubular reabsorption, a substance moves from the filtrate into the distal convoluted tubules to the peritubular capillaries. Active transport results in sodium, potassium, and glucose reabsorption. The presence of ADH causes water reabsorption.
Step 3: In tubular secretion, a substance moves from the peritubular capillaries into the tubular filtrate. Peritubular capillaries then secrete ammonia and hydrogen into the distal tubules via active transport.

Page 171

Mind sprints

1. Water
2. Potassium
3. Sulfates
4. Uric acid
5. Urobilinogen
6. Sodium
7. Calcium
8. Phosphates
9. Ammonium ions
10. Leukocytes
11. Chloride
12. Magnesium
13. Bicarbonates
14. Creatinine
15. RBCs

Page 172

You make the call

1. Low blood volume and increased serum osmolality are sensed by the hypothalamus, which signals the pituitary gland.
2. The pituitary gland secretes ADH into the bloodstream.
3. ADH causes the kidneys to retain water.
4. Water retention boosts blood volume and decreases serum osmolality.

Page 173

You make the call

1. Blood flow to the glomerulus drops.
2. Juxtaglomerular cells secrete renin into the bloodstream.
3. Renin travels to the liver.
4. Renin converts angiotensinogen in the liver to angiotensin I.
5. Angiotensin I travels to the lungs.
6. Angiotensin I is converted in the lungs into angiotensin II.
7. Angiotensin II travels to the adrenal glands.
8. Angiotensin II stimulates the adrenal glands to produce aldosterone.

■ Page 174
Starting lineup

> 1. Renin is secreted by the kidneys and is circulated in blood.

> 2. Renin causes the formation of angiotensin I.

> 3. Angiotensin I is converted into angiotensin II.

> 4. Angiotensin II constricts the arterioles.

> 5. Blood pressure rises.

■ Chapter 14

■ Page 179
Batter's box

1. balance, 2. balance, 3. metabolic, 4. homeostasis, 5. fluid, 6. skin, 7. urinary, 8. 1.5 L, 9. 2.6 L, 10. 800 ml, 11. 300 ml, 12. equal, 13. semipermeable, 14. diffusion, 15. active transport, 16. adenosine triphosphate

■ Page 180
Cross-training

			¹T	²A	L	K	A	L	³I	N	⁴E
			H						O		L
		⁵A	N	I	O	N	⁶S		N		E
			R				O		S		C
			S		L		⁷P				T
			T		U		H	⁸O			R
		⁹C	A	T	I	O	N	S			O
	¹⁰H		E		S		S	M			L
	Y				S		P	O			Y
¹¹A	C	I	D				H	S			T
	R						A	I			E
	¹²H	O	M	E	O	S	T	A	S	I	S
	G						E				
¹³P	R	O	T	E	I	N					
	N										

■ Page 181
Match point

1. C, 2. B, 3. D, 4. A

Power stretch

Isotonic: C, E
Hypotonic: A, D
Hypertonic: B, F

■ Page 182
Interval training

1. Sodium
2. 3.6 to 5 mEq/L
3. Calcium
4. 40 mEq/L
5. 4 mEq/L
6. Bicarbonate
7. Phosphate
8. 1 to 1.5 mEq/L

■ Page 183
Starting lineup

> 1. Serum osmolality drops to less than 280 mOsm/kg.

> 2. Thirst decreases.

> 3. ADH release is suppressed.

> 4. Renal water excretion increases.

> 5. Serum osmolality normalizes.

Power stretch

Sodium bicarbonate–carbonic acid: A, E, G
Phosphate: C, F
Protein: B, D, H

■ Page 184
Hit or miss

1. False. A rise in the carbon dioxide content of arterial blood or a decrease in blood pH stimulates the respiratory system, causing hyperventilation.
2. True.
3. True.
4. False. Recovery and formation of bicarbonate in the kidneys depend on hydrogen ion secretion by the renal tubules in exchange for sodium ions.
5. True.
6. False. The other factor influencing the rate of bicarbonate formation by renal tubular epithelial cells is the potassium content of the tubular cells.

■ Page 185
Team colors

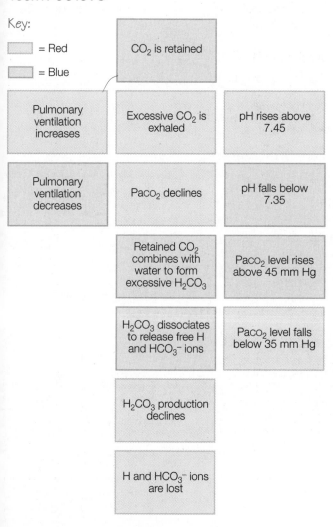

Key:

☐ = Red

☐ = Blue

CO_2 is retained

Pulmonary ventilation increases | Excessive CO_2 is exhaled | pH rises above 7.45

Pulmonary ventilation decreases | $Paco_2$ declines | pH falls below 7.35

Retained CO_2 combines with water to form excessive H_2CO_3 | $Paco_2$ level rises above 45 mm Hg

H_2CO_3 dissociates to release free H and HCO_3^- ions | $Paco_2$ level falls below 35 mm Hg

H_2CO_3 production declines

H and HCO_3^- ions are lost

■ Page 186
Starting lineup
Metabolic alkalosis

1. Chemical buffers bind with accumulated bicarbonate ions.

2. Excess bicarbonate elevates serum pH levels above 7.45, leading to decreased respiratory rate.

3. $Paco_2$ increases.

4. Bicarbonate that can't be reabsorbed by the renal glomeruli is excreted in urine.

5. Kidneys excrete excess sodium ions, water, and bicarbonate to maintain electrochemical balance.

6. Hydrogen ions diffuse out of cells, and potassium ions move into cells.

7. Calcium ionization decreases, allowing sodium ions to overstimulate nerve cells.

Metabolic acidosis

1. Chemical buffers bind with accumulated hydrogen ions.

2. Excess hydrogen decreases pH levels below 7.35, leading to increased respiratory rate.

3. $Paco_2$ decreases.

4. Kidneys secrete excess hydrogen ions into the renal tubules, which are then excreted in urine.

5. Hydrogen ions diffuse into cells, and potassium ions move into the blood.

6. Excess hydrogen ions lead to reduced excitability of nerve cells, causing central nervous system depression.

Chapter 15

■ Page 192

Cross-training

					¹S					²M		
	³A				P					Y		
⁴E	N	D	O	M	E	T	R	I	U	M	O	
	D				R		⁵A	M	P	U	L	L	A
	R		⁵A	M	P	U	L	L	A		E		
	O		A		T		⁶P				T		
	G		T		O	⁶P		⁷V	A	G	I	N	A
	E		O		E	N					U		
	N	⁸S	G										
	S	C	⁹E	P	I	D	I	D	Y	M	I	S	
		R	N	S									
		O	E										
¹⁰T	E	S	T	O	S	T	E	R	O	N	E		
		U	I										
		M		¹¹S	E	M	E	N					

■ Page 193

Finish line

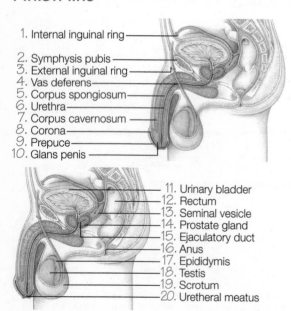

1. Internal inguinal ring
2. Symphysis pubis
3. External inguinal ring
4. Vas deferens
5. Corpus spongiosum
6. Urethra
7. Corpus cavernosum
8. Corona
9. Prepuce
10. Glans penis

11. Urinary bladder
12. Rectum
13. Seminal vesicle
14. Prostate gland
15. Ejaculatory duct
16. Anus
17. Epididymis
18. Testis
19. Scrotum
20. Uretheral meatus

■ Page 194

Batter's box

1. erectile tissue, 2. heavy fibrous tissue, 3. corpora cavernosa, 4. corpus spongiosum, 5. glans penis, 6. corona, 7. urethral meatus, 8. internal pudendal, 9. internal iliac vein

Match point

1. C, 2. B, 3. F, 4. E, 5. D, 6. A

■ Page 195

Train your brain

The male reproductive duct system conveys sperm from the testes to the ejaculatory ducts near the bladder.

Jumble gym

1. Epididymis, 2. Vas deferens, 3. Ampulla, 4. Urethra
Answer: Septum

■ Page 196

Power stretch

Seminal vesicles: B, E
Bulbourethral glands: D, F
Prostate gland: A, C, G

Circuit training

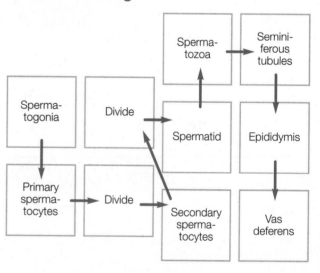

■ Page 197

Strike out

1. ~~Sperm formation begins at birth, although sperm don't mature until a male reaches puberty.~~ Sperm formation, or spermatogenesis, begins when a male reaches puberty and continues throughout life.
3. ~~Both spermatogonia and primary spermatocytes contain 46 sex chromosomes.~~ Both spermatogonia and primary spermatocytes contain 46 chromosomes; however, these consist of 44 autosomes and two sex chromosomes (X and Y).
6. ~~Each secondary spermatocyte also contains both an X and a Y chromosome.~~ Secondary spermatocytes contain either an X or a Y chromosome, not both.
9. ~~Newly mature sperm are stored in the epididymis until emission.~~ While a small number of sperm can be stored in the epididymis, most are stored in the vas deferens.
10. ~~Sperm cells lose their potency after 4 days.~~ Sperm cells retain their potency in storage for many weeks. After ejaculation, sperm can survive for up to 4 days in the female reproductive tract.

■ Page 198

Hit or miss

1. False. Androgens are produced in the testes and the adrenal glands.
2. False. Leydig's cells, located in the testes between the seminiferous tubules, secrete testosterone.
3. True.
4. True.
5. False. Testosterone secretion begins approximately 2 months after conception.
6. True.
7. True.
8. False. A male usually achieves full physical maturity by age 20.
9. True.
10. False. The prostate gland enlarges with age while its secretions diminish.

■ Page 199

Match point

1. D, 2. G, 3. A, 4. H, 5. L, 6. F, 7. J, 8. B, 9. E, 10. I, 11. K, 12. C

Finish line

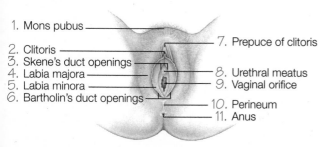

1. Mons pubus
2. Clitoris
3. Skene's duct openings
4. Labia majora
5. Labia minora
6. Bartholin's duct openings
7. Prepuce of clitoris
8. Urethral meatus
9. Vaginal orifice
10. Perineum
11. Anus

■ Page 200

Finish line

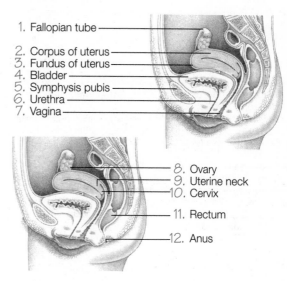

1. Fallopian tube
2. Corpus of uterus
3. Fundus of uterus
4. Bladder
5. Symphysis pubis
6. Urethra
7. Vagina
8. Ovary
9. Uterine neck
10. Cervix
11. Rectum
12. Anus

■ Page 201

Hit or miss

1. False. The vaginal wall has three tissue layers: epithelial tissue, loose connective tissue, and muscle tissue.
2. True.
3. False. The cervix connects the uterus to the vaginal vault. The perineum is located between the lower vagina and anal canal.
4. True.
5. True.
6. False. The vagina has three main functions: accommodating the penis during coitus, channeling blood discharged from the uterus, and serving as a birth canal.
7. False. Childbirth permanently alters the cervix, transforming it from a round opening about 3mm in diameter to a transverse slit with irregular edges.

Match point

1. D, 2. C, 3. G, 4. A, 5. F, 6. H, 7. E, 8. B

■ Page 202

Jumble gym

1. Fundus, 2. Fundibulum, 3. Hypogastric, 4. Fimbriae,
5. Graafian follicles

Answer: Fallopian tube

Strike out

2. ~~Ovaries are round, smooth, and pink at puberty.~~ Ovaries are round, smooth and pink at birth; they grow larger, flatten, and turn grayish by puberty.
3. ~~At menopause, ovaries take on an almond shape and a rough, pitted surface.~~ During childbearing years, ovaries take on an almond shape and a rough, pitted surface. At menopause, the ovaries shrink and turn white.
4. ~~At puberty, ovaries begin to produce graafian follicles.~~ At birth, each ovary contains approximately 500,000 graafian follicles.
7. ~~The ovaries' only function is to produce mature ova.~~ The ovaries' main function is to produce mature ova. However, they also produce estrogen, progesterone, and a small amount of androgens.

■ Page 203

Batter's box

1. mammary, 2. accessory, 3. milk, 4. lobes, 5. acini, 6. lactiferous, 7. ampullae, 8. areola, 9. Cooper's ligaments, 10. Montgomery's tubercles, 11. sebum

Hit or miss

1. False. The female reproductive cycle usually lasts 28 days, although it may range from 22 to 34 days.
2. False. Ovulatory, hormonal, and endometrial changes occur simultaneously.
3. True.
4. False. The follicular phase is the phase of the cycle during which a follicle develops. The luteal phase follows ovulation.
5. True.
6. True.

■ Page 204

Finish line

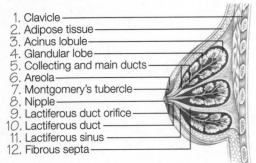

1. Clavicle
2. Adipose tissue
3. Acinus lobule
4. Glandular lobe
5. Collecting and main ducts
6. Areola
7. Montgomery's tubercle
8. Nipple
9. Lactiferous duct orifice
10. Lactiferous duct
11. Lactiferous sinus
12. Fibrous septa

■ Page 205

Circuit training

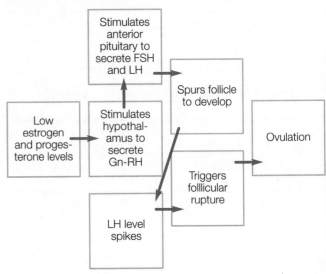

Batter's box

1. 7, 2. 5, 3. functional, 4. basal, 5. proliferative, 6. estrogen, 7. 14, 8. progesterone, 9. secretory, 10. corpus luteum, 11. progesterone

■ Page 206

Strike out

1. ~~Ovulation continues until the time of menopause.~~ Ovulation usually stops 1 to 2 years before menopause.
2. ~~As the ovaries reach the end of their productive cycle, they become unresponsive to estrogen stimulation.~~ As the ovaries reach the end of their productive cycle, they become unresponsive to gonadotropic stimulation.
5. ~~The pH of vaginal secretions decreases, making the risk of vaginal infections minimal.~~ The pH of vaginal secretions increases, making the vaginal environment more alkaline and more prone to infection.

Train your brain

A woman reaches menopause after menses are absent for one year.

■ Chapter 16

■ Page 211

Batter's box

1. fertilization, 2. spermatozoon, 3. ovum, 4. vagina, 5. canal, 6. mucus, 7. menstrual, 8. uterine cavity, 9. endometrium, 10. gestation, 11. zygote, 12. fetus, 13. labor

■ Page 212

Power stretch

Pre-embryonic: B, D, F, K
Embryonic: C, G, H, J
Fetal: A, E, I

You make the call

1. The spermatozoon, which has a covering called the *acrosome,* approaches the ovum.
2. The acrosome develops small perforations through which it releases enzymes necessary for the sperm to penetrate the protective layers of the ovum before fertilization.
3. The spermatozoon then penetrates the zona pellucida (the inner membrane of the ovum). This triggers the ovum's second meiotic division (following meiosis), making the zona pellucida impenetrable to other spermatozoa.
4. After the spermatozoon penetrates the ovum, its nucleus is released into the ovum, its tail degenerates, and its head enlarges and fuses with the ovum's nucleus. This fusion provides the fertilized ovum, called a *zygote,* with 46 chromosomes.

■ Page 213

Cross-training

■ Page 214

Starting lineup

First stage

| 1. Fetus begins descent. |
| 2. Contractiions cause cervical effacement. |
| 3. Full cervical dilation occurs. |

Second stage

| 4. Amniotic sac ruptures. |
| 5. Contractions increase in frequency and intensity. |
| 6. Fetus undergoes cardinal movements of labor. |
| 7. Fetal head is delivered. |
| 8. Fetal shoulders are delivered. |
| 9. Fetus is delivered. |

Third stage

| 10. Uterus continues to contract and shrinks. |
| 11. Placenta separates from the uterus. |
| 12. Placenta is delivered. |

Notes

Notes

Notes

Notes

Notes